Music and Mathematics

From Pythagoras to Fractals

Robert Fludd's *Temple of music* (1618), a complex amalgam of musical references, show Pythagoras entering the blacksmith's forge in the basement. The numbers displayed above that scene testify to the Pythagorean relationship between numbers and harmony.

Music and Mathematics

From Pythagoras to Fractals

Edited by

John Fauvel

Raymond Flood

and

Robin Wilson

OXFORD

UNIVERSITY PRESS

OXFORD

UNIVERSITY PRESS

Great Clarendon Street, Oxford OX2 6DP

Oxford University Press is a department of the University of Oxford.
It furthers the University's objective of excellence in research, scholarship,
and education by publishing worldwide in

Oxford New York

Auckland Bangkok Buenos Aires Cape Town Chennai
Dar es Salaam Delhi Hong Kong Istanbul Karachi Kolkata
Kuala Lumpur Madrid Melbourne Mexico City Mumbai Nairobi
São Paulo Shanghai Taipei Tokyo Toronto

Oxford is a registered trade mark of Oxford University Press
in the UK and in certain other countries

Published in the United States
by Oxford University Press Inc., New York

© John Fauvel, Raymond Flood, and Robin Wilson, 2003

A catalogue record for this title is available from the British Library

Library of Congress Cataloging in Publication Data
Data available
ISBN 0 19 851187 6

10 9 8 7 6 5 4 3 2

Typeset by Newgen Imaging Systems (P) Ltd., Chennai, India
Printed and bound in Great Britain by
Antony Rowe Ltd., Chippenham, Wiltshire

Preface

From ancient Greek times, music has been seen as a mathematical art. Some of the physical, theoretical, cosmological, physiological, acoustic, compositional, analytical and other implications of the relationship are explored in this book, which is suitable both for musical mathematicians and for musicians interested in mathematics, as well as for the general reader and listener.

In a collection of wide-ranging papers, with full use of illustrative material, leading scholars join in demonstrating and analysing the continued vitality and vigour of the traditions arising from the ancient beliefs that music and mathematics are fundamentally sister sciences. This particular relationship is one that has long been of deep fascination to many people, and yet there has been no book addressing these issues with the breadth and multi-focused approach offered here.

This volume is devoted to the memory of John Fauvel, Neil Bibby, Charles Taylor and Robert Sherlaw Johnson, whose untimely deaths occurred while this book was being completed.

Raymond Flood
Robin Wilson

February 2003

CONTENTS

Music and mathematics: an overview   1
Susan Wollenberg

And so they have handed down to us clear knowledge of the speed of the heavenly bodies and their risings and settings, of geometry, numbers and, not least, of the science of music. For these sciences seem to be related.

ARCHYTAS OF TARENTUM, EARLY FOURTH CENTURY BC

'We must maintain the principle we laid down when dealing with astronomy, that our pupils must not leave their studies incomplete or stop short of the final objective. They can do this just as much in harmonics as they could in astronomy, by wasting their time on measuring audible concords and notes.'

'Lord, yes, and pretty silly they look', he said. 'They talk about "intervals" of sound, and listen as carefully as if they were trying to hear a conversation next door. And some say they can distinguish a note between two others, which gives them a minimum unit of measurement, while others maintain that there's no difference between the notes in question. They are all using their ears instead of their minds.'

'You mean those people who torment catgut, and try to wring the truth out of it by twisting it on pegs.'

PLATO, FOURTH CENTURY BC

The Pythagoreans considered all mathematical science to be divided into four parts: one half they marked off as concerned with quantity, the other half with magnitude; and each of these they posited as twofold. A quantity can be considered in regard to its character by itself or in its relation to another quantity, magnitudes as either stationary or in motion. Arithmetic, then, studies quantity as such, music the relations between quantities, geometry magnitude at rest, spherics magnitude inherently moving.

PROCLUS, FIFTH CENTURY

This science [mathematics] is the easiest. This is clearly proved by the fact that mathematics is not beyond the intellectual grasp of any one. For the people at large and those wholly illiterate know how to draw figures and compute and sing, all of which are mathematical operations.

ROGER BACON, c.1265

I do present you with a man of mine,
Cunning in music and in mathematics,
To instruct her fully in those sciences,
Whereof, I know, she is not ignorant.

WILLIAM SHAKESPEARE, 1594

May not Music be described as the Mathematic of Sense, Mathematic as the Music of reason? The soul of each the same! Thus the musician feels Mathematic, the mathematician thinks Music,—Music the dream, Mathematic the working life,—each to receive its consummation from the other.

JAMES JOSEPH SYLVESTER, 1865

Mathematics and music, the most sharply contrasted fields of intellectual activity which can be found, and yet related, supporting each other, as if to show forth the secret connection which ties together all the activities of our mind . . .

H. VON HELMHOLTZ, 1884

Music is the arithmetic of sounds as optics is the geometry of light.

CLAUDE DEBUSSY, c.1900

Quite suddenly a young violinist appeared on a balcony above the courtyard. There was a hush as, high above us, he struck up the first great D minor chords of Bach's Chaconne. All at once, and with utter certainty, I had found my link with the centre . . . The clear phrases of the Chaconne touched me like a cool wind, breaking through the mist and revealing the towering structures beyond. There had always been a path to the central order in the language of music, . . . today no less than in Plato's day and in Bach's. That I now knew from my own experience.

WALTER HEISENBERG, 1971

When Professor Spitta, the great expert on Bach, explained to [Ethel Voynich (Lily Boole)] that in tuning, the third and fourth notes of the octave had to be just a little off or otherwise the octave would not fit, she suddenly "began to hate God and to despise the Almighty Creator of all things visible and invisible who couldn't make even eight notes fit", and she remained devoutly atheistic for the rest of her days. When Anne Freemantle told her many years later that Einstein had shown that it was only in our space-time continuum that the octave does not fit, the ninety-six year old Voynich replied reflectively, "Yes, perhaps I was a bit hasty."

DES MacHALE, 1985

Margarita Philosophica.

Music and mathematics: an overview

Susan Wollenberg

Mathematics and music have traditionally been closely connected. The seventeenth century has been seen by historians as a crucial turning-point, when music was changing from science to art, and science was moving from theoretical to practical. Many connections between science and music can be traced for this period. In the nineteenth and twentieth centuries, the development of the science of music and of mathematical approaches to composition further extended the connections between the two fields. Essentially, the essays in this book share the concern of commentators throughout the ages with the investigation of the power of music.

Musicke I here call that *Science*, which of the Greeks is called Harmonie...
Musicke is a Mathematical Science, which teacheth, by sense and reason, perfectly to judge, and order the diversities of soundes hye and low.

JOHN DEE (1570)

The invitation to write an introduction to this collection offered a welcome opportunity to reflect on some of the historical, scientific, and artistic approaches that have been developed in the linking of mathematics and music. The two have traditionally been so closely connected that it is their separation that elicits surprise. During the late sixteenth and early seventeenth centuries when music began to be recognized more as an art and to be treated pedagogically as language and analysed in expressive terms, it might have been expected to lose thereby some of its scientific connotations; yet in fact the science of music went on to develop with renewed impetus.

This introduction sets out to explore, via a variety of texts, some of the many historical and compositional manifestations of the links between mathematics and music. (This endeavour cannot be other than selective: the field is vast, ranging from ancient theory and early developments in structure such as those of the medieval motet, to the new ideas of post-tonal music and experimental musical techniques explored over the past century.) In what follows, the field is viewed particularly from the perspective of a music historian with a special interest in the history of music in its educational dimension.

In the traditional arrangement of knowledge and teaching in universities, music was one of the seven liberal arts, along with the other quadrivium subjects of arithmetic, geometry and astronomy. This woodcut dates from 1504.

Aspects of notation and content

In contemplating the two disciplines, mathematics and music (and taking music here essentially to mean the Western 'Classical' tradition), it is clear to the observer from the outset that they share some of their most basic properties. Both are primarily (although not exclusively) dependent on a specialized system of notation within which they are first encoded by those who write them, and then decoded by those who read (and, in the case of music, perform) them. Their notations are both ancient and modern, rooted in many centuries of usage while at the same time incorporating fresh developments and newly-contrived systems to accommodate the changing patterns of mathematical and musical thought.

Musical notation can be traced back to the ancient Greek alphabet system. A series of significant stages came in the development of notations within both the Western and Eastern churches during the medieval period. In the eleventh to thirteenth centuries more precise schemes were codified, including Guido d'Arezzo's new method of staff notation and the incorporation of rhythmic indications. By the time of the late sixteenth and the seventeenth centuries, most of the essential features of musical notation as it is commonly understood today were in place within a centrally established tradition. Subsequent additions were mainly in the nature of surface detail, although of considerable importance, as with the expanded range of performance instructions in the nineteenth century. The twentieth century, with its emphasis on experimental music, saw a precipitate rise in new forms of notation. In a comparable way, mathematical notation has developed over a period of at least 2500 years and, in doing so, has inevitably drawn from various traditions and sources.

In music, the relationship between notation and the content it conveys is sometimes more complex than might at first appear. Notation has not invariably fulfilled the role merely of servant to content. While it is generally true that notational schemes evolved in response to the demands posed by new ideas and new ways of thinking, it is also possible that experiments in notation may have been closely fused with the development of such ideas, or may even have preceded—and inspired— their creation. In mathematics, too, the relationship has subtle nuances. Notation developed in one context could prove extremely useful in another (seemingly quite different) context. (A well-known example of this is the use of tensor notation in general relativity.) In one notable case, notation formed part of the focus of a professional dispute, when a prolonged feud developed between Newton and Leibniz as to which of them invented the differential calculus, together with the different notation used by each.

In the course of their history, mathematics and music have been brought together in some curious ways. The Fantasy Machine demonstrated in 1753 by the German mathematician Johann Friedrich Unger to the Berlin Academy of Sciences, under Leonhard Euler's presidency, was designed to preserve musical improvisations; in the words of an English inventor, the Revd John Creed, on whose behalf a similar idea was presented to the Royal Society in London in 1747, with this device the 'most transient Graces' could be 'mathematically delineated'. Unger claimed to have had the idea as early as 1745, although Charles Burney (in his essay 'A machine for recording music') attributed priority of invention to Creed. Although it aroused considerable interest and support among the intelligentsia, and 'was tried out by several well-known musicians' in the mid-eighteenth century, the machine was ultimately not a success.

Music as science: the historical dimension

Throughout the history of mathematical science, mathematicians have felt the lure of music as a subject of scientific investigation; an intricate network of speculative and experimental ideas has resulted. Taking a historical view, Penelope Gouk has voiced her concern that such terms as 'mathematical sciences' are 'routinely used as essentially unproblematical categories which are self-evidently distinct from the arts and humanities...Since music is today regarded as an art rather than a science, it is hardly surprising that the topic should be disregarded by historians of science'. Her book remedies this situation with resounding success, inviting a reconsideration of the way joint histories are told.

Within the scope of a work based primarily on seventeenth-century England, Gouk's references range from Pythagoras (in particular, Pythagorean tuning and the doctrine of universal harmony that 'formed the basis of the mathematical sciences') to René Descartes ('the arithmetical foundation of consonance') and beyond. Descartes' *Compendium* (1618) was translated as *Renatus Des-Cartes excellent compendium of musick and animadversions of the author* (1653) by the English mathematician William Brouncker. Brouncker himself was 'the first English mathematician to apply logarithms (invented *c.*1614) to the musical division'. Thus he entered into a scientific dialogue with the work of Descartes, contesting the latter's findings.

At the period when music was changing from science to art (retaining a foot in both camps), science itself was moving from theoretical to practical. The seventeenth century has been seen by historians as a crucial turning-point, with the emergence of a 'recognizable scientific community' and the institutionalization of science. The founding of the Royal Society of London in 1660 formed a key point in the development

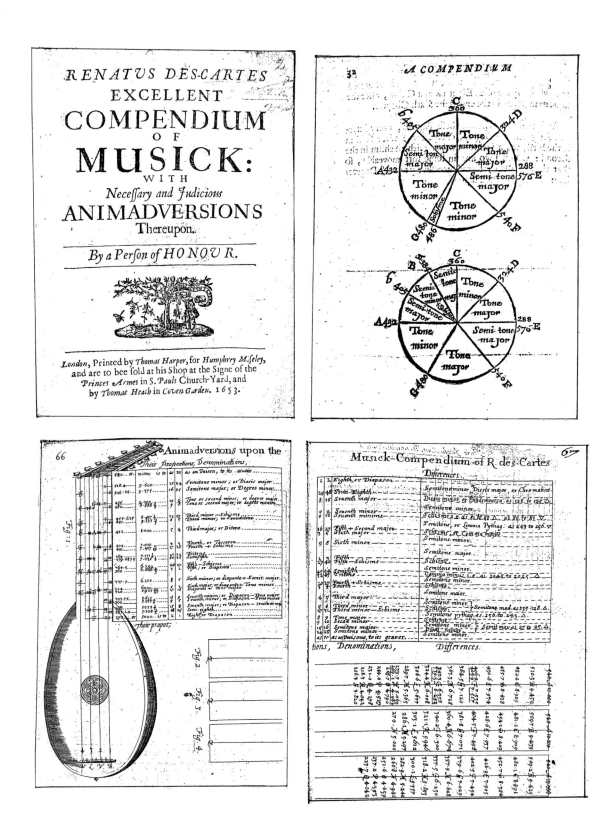

Four pages from Lord Brouncker's translation, *Renatus Des-Cartes excellent compendium of musick.*

of modern scientific enquiry. The leading scientific thinkers who gathered under its auspices focused some of their attention on music. Gouk notes that 'the [Royal] Society's most overt interest in musical subjects occurred within the presidencies of Brouncker and Moray, both of whom . . . were competent musicians and keen patrons of music'. Practitioners of both mathematics and music could learn much from each other's work.

Ideas such as those of musical tunings were constantly subject to review in the light of new theories. Musical issues occupied a central, not peripheral, position in science: 'the conceptual problems involved in the division of musical space were among the most important challenges faced by seventeenth-century mathematicians and natural philosophers'. As Gouk observes, Newton in the mid-1660s 'learned all that had been developed by modern mathematicians such as Descartes, Oughtred and Wallis' regarding the musical scale, and especially the division of the scale, and 'rapidly went beyond them in his own studies'. Important discoveries of this period generally included the observation that 'pitch can be identified with frequency'. The seventeenth century saw the beginnings of modern acoustical science: the new science of sound.

The work of Mersenne has also been seen as representing 'a significant milestone in the emergence of modern science, just like the musical laws that he established'. Mersenne's writings—notably, his *Harmonie universelle* (1636) and *Harmonicorum libri* (see Chapter 2)— became available in England. (Gouk notes 'how rapidly Mersenne's work on musical acoustics was assimilated in England'.) Mersenne's belief that 'the universe was constructed according to harmonic principles expressible through mathematical laws' provided an impetus

The legend of Pythagoras's discovery of a relation between musical notes and hammer weights, as portrayed in a twelfth-century manuscript.

for mathematicians such as Newton. Personal contacts and correspondence among scientists further created and consolidated intellectual connections at this period.

Educationally, the influential tradition of Boethius (c.480–524), casting a long shadow over the following centuries, and based in its turn on Pythagoras and Plato, aligned music with arithmetic, astronomy and geometry in the *quadrivium*, while grammar, rhetoric and logic formed the language-based *trivium*. When the seven Gresham Professorships were founded in the City of London in 1596 to provide free adult education, their subjects included music, 'physic', geometry and astronomy. At the opposite end of the educational spectrum, Henry Savile's 1619 foundation of the University Chair in mathematics at Oxford University included, in its stipulations of the new professor's duties, that he was to expound on 'canonics, or music' as one of the quadrivial disciplines. Music was taught at the universities as a science, while it was examined (in the form of the B.Mus. and D.Mus. degrees) as an art, by means of the submission of a composition.

Among the spate of Professorships endowed during the early decades of the seventeenth century, William Heather's founding of the Chair in Music at Oxford (1627) recognized this duality with its provision for the regular practice of music as well as lectures on the science of music. In doing so, Heather reflected Thomas Morley's two-fold division in his *Plaine and easie introduction to music* (1597):

Speculative is that kinde of musicke which by Mathematical helpes, seeketh out the causes, properties, and natures of soundes... content with the onlie contemplation of the Art, *Practical* is that which teacheth al that may be known in songs, eyther for the understanding of other mens, or making of one's owne...

Scientific musical enquiry, analytical listening (or listening with understanding), and the art of composition, are all equally acknowledged as valid activities here.

In his pioneering lectures published in 1831, the Heather Professor of Music, William Crotch was in no doubt as to music's position in the scheme of things: from the outset of his 'Chap. 1: Introductory', he asserted that 'Music is both an art and a science'. Crotch followed this opening gambit with a long and particularly apposite quotation from the work of Sir William Jones:

Music... belongs, as a science, to an interesting part of natural philosophy, which, by mathematical deductions... explains the causes and properties of sound... but, considered as an art, it combines the sounds which philosophy distinguishes, in such a manner as to gratify our ears, or affect our imaginations; or, by uniting both objects, to captivate the fancy, while it pleases the sense; and speaking, as it were, the language of nature, to raise corresponding ideas and connections in the mind of the hearer. It then, and then only, becomes fine art, allied very nearly to poetry, painting, and rhetoric...

William Heather, founder of the Heather Chair in Music at Oxford University, and William Crotch, a later holder of that Chair.

While Crotch went on to state that 'The science of music will not constitute the subject of the present work', he nevertheless used this as a device to launch into a discussion of the merits of such an enquiry, strongly recommending 'the study . . . of the science of music . . . to every lover of the art', and pursuing some of its ramifications at considerable length before concluding that 'enough . . . has now been said, to induce the lover of music to study the science, which, it will be remembered, is not the proper subject of this work'. After some ten pages of discussion the reader might well have forgotten this assertion, or be inclined to question it; and it is clear that Crotch felt it inappropriate to offer to the public a didactic treatise on music without paying any consideration to its scientific dimension, even though his primary purpose in presenting these lectures was an aesthetic one ('being the improvement of taste').

Some social and educational connections

In the more informal sphere, the history of cultural life is liberally scattered with examples of musical mathematicians and scientists. The group of intellectuals and artists to which C. P. E. Bach belonged in eighteenth-century Hamburg, and which included J. J. C. Bode (translator of, among other works, Sterne's *A sentimental journey*), met regularly at the house of the mathematician J. G. Büsch; 'many were keen amateur musicians, including Bode who played the cello in the regular music-making at Büsch's house'. C. P. E. Bach's biographer, Hans-Günter Ottenberg, has written of 'the friendly atmosphere and liberal exchange of ideas which took place at the home of the mathematician Johann Georg Büsch . . .', quoting Reichardt's description of these gatherings, which evidently possessed a certain cachet: 'not everyone was admitted to the inner circle which would not infrequently assemble for a pleasant evening's entertainment apart from the wider academic community'. Ottenberg stresses that C. P. E. Bach was 'one of Büsch's closer acquaintances'.

In nineteenth-century Oxford, Hubert Parry, as an undergraduate, frequented the home of Professor Donkin (Savilian Professor of Astronomy) where the Donkins—a highly musical family altogether—held chamber-music gatherings. For Parry these occasions and the opportunities they provided, both for getting to know the chamber music repertoire and for composing his own efforts in the genre, were enormously stimulating. The Donkins were influential figures in Oxford's musical life during the second half of the nineteenth century.

It was in this period, too, that the academic status of music, in the shape of the Oxford musical degrees, acquired greater weight. The succession of Heather Professors of Music at Oxford and their assistants voiced their hopes for the development of the subject within the

Hubert Parry, as an Oxford undergraduate, and William Donkin, Savilian Professor of Astronomy.

University, including serious consideration given to the science of music; for example, the set texts for the D.Mus. at Oxford included Helmholtz (see Chapter 5), and others, on acoustics. The evidence presented by Sir Frederick Ouseley (then Heather Professor) to the University of Oxford Commission in 1877 included a 'Proposal for establishing a Laboratory of Acoustics' (apparently this plan was never realised); Ouseley envisaged that such a laboratory 'might work in with the scientific side of a school of technical music' and would have 'more direct relations with the school of physics in the University'.

Holders of music degrees from Oxford during this period (qualifications that were considerably coveted in the musical profession) did not all follow primarily musical careers; William Pole FRS (b.1814, B.Mus. 1860, D.Mus. 1867) was Professor of Civil Engineering at University College, London, as well as organist of St Mark's, North Audley Street. Among those who took the B.Mus. at Oxford, in addition to the ordinary BA, was J. Barclay Thompson of Christ Church (B.Mus. 1868), who became University Reader in Anatomy. More recent scientist-musicians have included the mathematically trained musicologist Roy Howat, whose work on the golden section in Ravel's music, among other topics, has attracted wide interest.

Mathematics and music: the compositional dimension

While music has fascinated mathematical scientists as a subject of enquiry, musicians have been attracted by the possibilities of incorporating mathematical science into their efforts, most notably in the fields of composition and analysis. The fundamental parameters of music—pitch, rhythm, part-writing, and so on—and the external ordering of musical units into a set, have lent themselves to systematic arrangement reflecting mathematical planning. Much has been written about the mathematical aspects of particular compositional techniques—for example Schoenberg's method of serialism (see Chapter 8)—and individual works have frequently been analysed in terms of their mathematical properties, among other aspects.

The possibilities of mathematical relationships not only within a single piece, but also between a number of pieces put together to form a set, are well documented. These sorts of schemes may be expressed in the findings of musical analysts, possibly by reconstructing notional systems of composition, and, further, by examining both the known and the speculative symbolic associations, as well as the mathematical ramifications, of such structural procedures. This is found most obviously in the case of number symbolism, which may be perceived as governing the musical relationships of an individual piece or a whole set of pieces.

Contrapuntal techniques in music have traditionally been treated mathematically and identified with qualities of rigour. Among the prime examples in these latter two categories—the compositional set, and rigorous counterpoint—must be counted the works of J. S. Bach, with their mirror canons and fugues, their ordering by number (as with the *Goldberg variations*), and their emphasis on combinatorial structures. At a distance of over 200 years, Paul Hindemith's cycle of fugues and interludes for piano, the *Ludus tonalis*, with its 'almost geometric design', its pairs of pieces mirroring each other (see Chapter 6), provides a modern echo of these contrapuntal ideas very much in the Bach tradition, as well as building on techniques developed in Hindemith's theoretical writings. It has been suggested, moreover, that Hindemith 'identified...closely with Kepler', whose life and work formed the subject of Hindemith's last full-length opera, *The harmony of the world* (1956–7).

'Scientific' music has not, however, always been appreciated by musical scientists. Christiaan Huygens, for instance, expressed a wish that composers 'would not seek what is the most artificial or the most difficult to invent, but what affects the ear most', professing not to care for 'accurately observed imitations called "fugues"', or for canons, and claiming that the artists who 'delight in them' misjudge the aim of music, 'which is to delight with sound that we perceive through the ears, not with the contemplation of art'. Huygens here articulated the tension between 'scientific' construction in musical composition, on the one hand, and music's expressive effect, on the other. The balance between these two aspects, and more widely between the scientific basis of the art of music and its aesthetic applications, has been a source of fascination for scholars, and indeed continues to be so, as the essays in this book serve collectively to demonstrate. Their shared concern is essentially the investigation of the power of music, which has preoccupied commentators throughout the ages, from antiquity to our own time.

PART I

Music and mathematics through history

CHAPTER I

Tuning and temperament: closing the spiral

Neil Bibby

In Ancient Greek times it was recognized that consonant musical sounds relate to simple number ratios. Nevertheless, in using this insight to construct a scale of notes for tuning an instrument, problems arise. These problems are especially noticeable when transposing tunes so that they can be played in different keys. A solution adopted in European music over the last few centuries has been to draw upon mathematics in a different way, and to devise an 'equally-tempered' scale.

Each musical note has a basic frequency (essentially, the number of times the sound pulsates in a given period of time): thus the note 'A', which you may hear the oboe play while an orchestra is tuning up, has a frequency of 440 Hz (cycles per second). Frequency enables us to talk about relationships between musical sounds. However, for purposes of comparing two notes, the actual frequency is less important than the *ratio* of their frequencies.

The structure of a musical scale is determined by the frequency ratios of the notes that form the scale. The choice of these ratios is ultimately governed by the degree of *consonance* between the notes. Consonance is both a psychological and a physical criterion: two notes are *consonant* if they sound 'pleasing' when played together. In physical terms this seems to occur when the frequency ratio of the two notes is a ratio of low integers: the simpler the ratio, the more consonant are the two notes.

Apart from the trivial case of a unison, for which the frequency ratio is $1:1$, the simplest case is the frequency ratio $2:1$. When two notes have this frequency ratio the interval between them is an *octave*: thus, for the oboe A, the next higher A has frequency 880 Hz. The origins of this interval may lie in pre-history, when the earliest attempts at group singing or chanting would have been in unison, or in octaves for mixed groups of adults, or men and children: the different vocal ranges of the participants would thus force the harmonic use of the octave instead of the unison. As a melodic interval the octave is not common, but three

Ancient harmonic discoveries are portrayed in this woodcut from Franchino Gafurio's *Theorica musice* (1492). Mathematical ratios are emphasized in the experiments attributed to Pythagoras.

popular twentieth-century American songs that start with a rising octave are *Somewhere over the rainbow, Singin' in the rain*, and *Bali Hai*.

This simple frequency relationship of 2:1, corresponding to two notes forming an octave, is the basis for the construction of any musical scale. Mathematically, the problem of constructing the scale is to determine an appropriate set of frequency ratios for the notes that lie in between. The number of these interpolated notes is arbitrary from a mathematical point of view. However, the frequency ratios of the intervening notes must satisfy the psychological/aesthetic criterion of consonance. Ultimately, as we shall see, the mathematical criterion of simplicity that underlies the early notion of consonance yields to other mathematical criteria. It turns out that the tolerance of the human ear, together with natural conditioning, enables the 'simplicity' criterion to be partially relaxed.

The Pythagorean scale

The oldest system of scale construction is that described as the *Pythagorean scale*. The system is much older than Pythagoras (*c*.550 BC), but his name is associated with the theoretical justification, in mathematical terms, of its construction. Legends have come down to us, through the late Roman writer Boethius among others, relating how Pythagoras 'discovered' this scale: they alleged that Pythagoras noted the harmonious relationships of the sounds produced by the hammers in a blacksmith's forge, and further investigations revealed that the masses of these hammers were, extraordinarily, in simple whole-number ratios to each other! From this claimed observation Pythagoras is supposed to have leapt to the realization that consonant sounds and simple number ratios are correlated—that ultimately music and mathematics share the same fundamental basis.

It is not difficult to construct a scale by following the Pythagorean insight. The strategy is to take any note and produce others related to

Detail from Robert Fludd's *Temple of music* (see Frontispiece), showing Pythagoras entering the blacksmith's forge.

it by simple whole-number ratios, in the confidence that on Pythagorean principles the resultant notes will sound consonant. The structure of such a scale is ultimately based on the simple frequency ratios of $2:1$ and $3:1$.

In the case of a plucked or bowed string, different notes may be produced depending on how the string vibrates, and this too seems to follow the Pythagorean observation. Consider a vibrating string sounding a note of frequency t.

The same string can also vibrate at twice the original frequency, sounding the note of frequency $2t$. The *interval* between the new and original notes is given by the ratio of the frequencies, $2t:t$ or $2:1$, an octave.

If the string were to vibrate with three times the original frequency, it would sound a note of frequency $3t$.

The interval between the notes of frequencies $3t$ and $2t$ is $3:2$, or $\frac{3}{2}$. Equivalently, the note an octave below $3t$ is $\frac{3}{2}t$, and the interval between the note with frequency t and this note is therefore $\frac{3}{2}$.

We now have a three-note scale $\{t, \frac{3}{2}t, 2t\}$. If we regard the note with frequency t as the note C, for example, with C' an octave higher, then this scale is

C	G	C'
t	$\frac{3}{2}t$	$2t$

This procedure has not only created a new note (G), but also a further new interval. Our previous interval, between C and G, is called a *perfect fifth* and the new interval between G and C' is called a *perfect fourth*. The ratio corresponding to the perfect fifth is $\frac{3}{2}$, as we have seen, while the perfect fourth has ratio $2t:\frac{3}{2}t$, or $\frac{4}{3}$.

We now have a method for generating yet more notes. If we lower the note C' by a perfect fifth, by dividing its frequency by $\frac{3}{2}$, we obtain the note F of frequency $\frac{4}{3}t$. It lies between C and G, and the resulting scale is

C	F	G	C'
t	$\frac{4}{3}t$	$\frac{3}{2}t$	$2t$

The process by which the scale is generated is thus essentially iterative: each new note yields a new interval with its nearest neighbour, and this interval can then be used to generate further new notes.

By continuing in this way, we obtain the interval between F and G. This is called the *major second*, or *whole tone*, and has ratio $\frac{3}{2}t : \frac{4}{3}t$, or $\frac{9}{8}$. This new interval in turn gives rise to a new note by simultaneously lowering both F and G by a perfect fourth: the new note, a whole tone above C, is D. We can now use the whole tone interval to fill in the gaps in the scale:

name of note	C	D	E	F	G	A	B	C'
frequency	$\frac{1}{1}t$	$\frac{9}{8}t$	$\frac{81}{64}t$	$\frac{4}{3}t$	$\frac{3}{2}t$	$\frac{27}{16}t$	$\frac{243}{128}t$	$\frac{2}{1}t$
interval		$\frac{9}{8}$	$\frac{9}{8}$	$\frac{256}{243}$	$\frac{9}{8}$	$\frac{9}{8}$	$\frac{9}{8}$	$\frac{256}{243}$

Each of the resulting 'narrow' intervals E to F and B to C is a *minor second*, or *semitone*, and has a ratio of $\frac{4}{3} : \frac{81}{64}$, which is $\frac{256}{243}$. In addition, several other new intervals appear, including the major third C to E, with ratio $\frac{81}{64}$, the major sixth C to A, with ratio $\frac{27}{16}$, and the major seventh C to B, with ratio $\frac{243}{128}$. We thus arrive at the Pythagorean scale, and we denote the resulting set of notes by **P**.

An alternative view is to regard the scale as being formed by a succession of perfect fifths, starting from C. In this view, we form the five notes that are successive fifths above C, and the note that is a perfect fifth below C. We then reassemble these into a single octave.

The result of this process is equivalent to our earlier one. In the resulting scale, successive notes are separated by an interval of a tone, with ratio $\frac{9}{8}$, or a semitone, with ratio $\frac{256}{243}$. The semitone is actually smaller than its name would suggest, because $\left(\frac{256}{243}\right)^2$ is less than $\frac{9}{8}$—so it is not a 'semi'-tone in any accurate sense! We shall see later that this leads to serious problems: for example, on a modern keyboard it *seems* as though twelve perfect fifths are equivalent to seven octaves. However, if the tuning is Pythagorean, this cannot possibly be the case, as we shall see later.

More generally, if we stick to octaves and perfect fifths, then only the numbers 2 and 3 (and their powers) can be involved in these ratio calculations. Thus, each note in the Pythagorean scale can be written simply as $2^p \cdot 3^q$, where p and q are integers: here, and from now on, we omit the factor t. The scale **P** can thus be represented as follows:

C	D	E	F	G	A	B	C'
1	$3^2/2^3$	$3^4/2^6$	$2^2/3$	$3/2$	$3^3/2^4$	$3^5/2^7$	2

Exploring further the way that the notes of the Pythagorean scale combine, however, we run into a problem. Suppose that we wish to find the note a major seventh above A ($3^3/2^4$): this note is $3^3/2^4 \times 3^5/2^7 = 3^8/2^{11}$. Lowering this by an octave, we get $3^8/2^{12}$, which must lie somewhere between G and A (since $3/2 < 3^8/2^{12} < 3^3/2^4$). This leads us to realize that the Pythagorean scale is not 'closed' under transposition, but the rules under which we have constructed the scale will lead to an indefinite number of new notes. This leads to problems if we want to construct a scale (in particular, a physically embodied scale such as a keyboard) that allows transposition of keys.

Transposition in the Pythagorean scale

We constructed the Pythagorean scale **P** by a succession of transpositions of the basic key note C: in each case we transposed up a fifth (multiplying its frequency by $\frac{3}{2}$) and where necessary took the resulting note down an octave (halving its frequency). A good way of seeing what is going on in the problematic issue which has just arisen, of an apparently indefinite number of new notes being produced, is to consider the effect of the same transpositions on the entire scale **P**. Does this lead to another Pythagorean scale, and are the same notes involved?

Let us build a new scale on the note G. To do this, we transpose the original Pythagorean scale **P** up by a fifth, and transpose down an octave when necessary. The resulting scale **P**^{1} includes most of the notes of **P** itself, as a result of the partial regularity of the distribution of the intervals between the original notes:

[tone-tone-semitone]-tone-[tone-tone-semitone].

However, there is a 'new' element, the note $3^6/2^9$: this note lies between the two existing notes F and G, since $2^2/3 < 3^6/2^9 < 3/2$. This new note is the familiar F sharp, written F$^\sharp$, and is required when we transpose from the scale of C to the scale of G. It does not lie symmetrically between F and G, however, since the interval $3^7/2^{11}$ between F and F$^\sharp$ is slightly greater than the interval $2^8/3^5$ between F$^\sharp$ and G.

	C		D		E	F	F$^\sharp$	G		A		B	C'
P	1		$\frac{3^2}{2^3}$		$\frac{3^4}{2^6}$	$\frac{2^2}{3}$		$\frac{3}{2}$		$\frac{3^3}{2^4}$		$\frac{3^5}{2^7}$	2
P^1	1		$\frac{3^2}{2^3}$		$\frac{3^4}{2^6}$		$\frac{3^6}{2^9}$	$\frac{3}{2}$		$\frac{3^3}{2^4}$		$\frac{3^5}{2^7}$	2

In a similar way, a new scale can be built on the note F. In this case we *divide* the frequencies of each note by $\frac{3}{2}$, and where necessary transpose up an octave. This new scale, which we may call **P$_1$**, again contains a 'rogue' element, with frequency $2^4/3^2$, which is the familiar B flat, written B$^\flat$, of the key of F. Again, this new note lies between two existing notes, A and B, since $3^3/2^4 < 2^4/3^2 < 3^5/2^7$, and again not symmetrically since $2^8/3^5$ is less than $3^7/2^{11}$: thus, the new note is less than the geometric mean of the two notes each side of it.

	C		D		E	F		G		A	B$^\flat$	B	C'
P	1		$\frac{3^2}{2^3}$		$\frac{3^4}{2^6}$	$\frac{2^2}{3}$		$\frac{3}{2}$		$\frac{3^3}{2^4}$		$\frac{3^5}{2^7}$	2
P$_1$	1		$\frac{3^2}{2^3}$		$\frac{3^4}{2^6}$	$\frac{2^2}{3}$		$\frac{3}{2}$		$\frac{3^3}{2^4}$	$\frac{2^4}{3^2}$		2

Continuing in this way, we successively generate a new note between a pair of old notes, with each new note being slightly higher or lower than the geometric mean of its neighbours. After six such transpositions in each direction, we arrive at the scales **P^6** and **P$_6$** opposite, in each row of which only one note (F or B, respectively) has survived from the original scale **P**.

The notes of the top row correspond to the key of F$^\sharp$ and those of the bottom row correspond to that of G$^\flat$. By comparing these two scales, we can see that all of the notes of the G$^\flat$ scale are slightly lower than those of the F$^\sharp$ scale. In particular, under the transposition into the key of F$^\sharp$, the original key note C has become $3^6/2^9$, while under its transposition into G$^\flat$ it has become $2^{10}/3^6$. The interval between these notes is $(3^6/2^9)/(2^{10}/3^6)$, which simplifies to $3^{12}/2^{19}$ or $1.01364\ldots$. This very small difference, called the *Pythagorean comma*, lies at the root of the contradictions inherent in the Pythagorean scale. Although 3^{12} and 2^{19} are very close, they are not the same.

Furthermore, no succession of fifths can form an exact number of octaves—for if it did, there would be integer solutions p and q to the equation $(\frac{3}{2})^p = 2^q$, or $3^p = 2^{p+q}$. This has no solutions, since no power of 3 can equal a power of 2 (apart from the zeroth power), a particular

	C	C$\sharp$	D	D$\sharp$	E	F	F$\sharp$	G	G$\sharp$	A	A$\sharp$	B	C'	
P^6		$\frac{3^7}{2^{11}}$		$\frac{3^9}{2^{14}}$		$\frac{3^{11}}{2^{17}}$	$\frac{3^6}{2^9}$		$\frac{3^8}{2^{12}}$		$\frac{3^{10}}{2^{15}}$	$\frac{3^5}{2^7}$		F$\sharp$
P^5		$\frac{3^7}{2^{11}}$		$\frac{3^9}{2^{14}}$	$\frac{3^4}{2^6}$		$\frac{3^6}{2^9}$		$\frac{3^8}{2^{12}}$		$\frac{3^{10}}{2^{15}}$	$\frac{3^5}{2^7}$		B
P^4		$\frac{3^7}{2^{11}}$		$\frac{3^9}{2^{14}}$	$\frac{3^4}{2^6}$		$\frac{3^6}{2^9}$		$\frac{3^8}{2^{12}}$	$\frac{3^3}{2^4}$		$\frac{3^5}{2^7}$		E
P^3		$\frac{3^7}{2^{11}}$	$\frac{3^2}{2^3}$		$\frac{3^4}{2^6}$		$\frac{3^6}{2^9}$		$\frac{3^8}{2^{12}}$	$\frac{3^3}{2^4}$		$\frac{3^5}{2^7}$		A
P^2		$\frac{3^7}{2^{11}}$	$\frac{3^2}{2^3}$		$\frac{3^4}{2^6}$		$\frac{3^6}{2^9}$	$\frac{3}{2}$		$\frac{3^3}{2^4}$		$\frac{3^5}{2^7}$		D
P^1	1		$\frac{3^2}{2^3}$		$\frac{3^4}{2^6}$		$\frac{3^6}{2^9}$	$\frac{3}{2}$		$\frac{3^3}{2^4}$		$\frac{3^5}{2^7}$	2	G
P	1		$\frac{3^2}{2^3}$		$\frac{3^4}{2^6}$	$\frac{2^2}{3}$		$\frac{3}{2}$		$\frac{3^3}{2^4}$		$\frac{3^5}{2^7}$	2	C
P$_1$	1		$\frac{3^2}{2^3}$		$\frac{3^4}{2^6}$	$\frac{2^2}{3}$		$\frac{3}{2}$		$\frac{3^3}{2^4}$	$\frac{2^4}{3^2}$		2	F
P$_2$	1		$\frac{3^2}{2^3}$	$\frac{2^5}{3^3}$		$\frac{2^2}{3}$		$\frac{3}{2}$		$\frac{3^3}{2^4}$	$\frac{2^4}{3^2}$		2	B$\flat$
P$_3$	1		$\frac{3^2}{2^3}$	$\frac{2^5}{3^3}$		$\frac{2^2}{3}$		$\frac{3}{2}$	$\frac{2^7}{3^4}$		$\frac{2^4}{3^2}$		2	E$\flat$
P$_4$	1	$\frac{2^8}{3^5}$		$\frac{2^5}{3^3}$		$\frac{2^2}{3}$		$\frac{3}{2}$	$\frac{2^7}{3^4}$		$\frac{2^4}{3^2}$		2	A$\flat$
P$_5$	1	$\frac{2^8}{3^5}$		$\frac{2^5}{3^3}$		$\frac{2^2}{3}$	$\frac{2^{10}}{3^6}$		$\frac{2^7}{3^4}$		$\frac{2^4}{3^2}$		2	D$\flat$
P$_6$		$\frac{2^8}{3^5}$		$\frac{2^5}{3^3}$		$\frac{2^2}{3}$	$\frac{2^{10}}{3^6}$		$\frac{2^7}{3^4}$		$\frac{2^4}{3^2}$	$\frac{2^{12}}{3^7}$		G$\flat$
	C	D$\flat$	D	E$\flat$	E	F	G$\flat$	G	A$\flat$	A	B$\flat$	B	C'	

Pythagorean scales.

case of a mathematical result (the uniqueness of prime factorization) known since the time of Euclid. However, the fact that 3^{12} is approximately equal to 2^{19} suggests that $p = 12$, $q = 7$ is an approximate solution, and that the 'difference' can be measured by the ratio $3^{12}/2^{19}$, the Pythagorean comma.

We are thus faced with the fact that there is no end to the process we have initiated: transposition up a fifth and transposition down a fifth take us on infinite journeys, ever generating new notes, even if some of these (as with G$\flat$ and F$\sharp$) are tantalisingly close. The journey can be thought of as traversing a spiral, starting from our set **P** (represented by C): for each $30°$ step clockwise we spiral outwards and transpose up a fifth, while for each $30°$ step anti-clockwise we spiral inwards and transpose down a fifth (see overleaf). Adjacent points on the same ray of the spiral differ by the Pythagorean comma.

Just intonation

Many of the intervals produced by the Pythagorean system are far from simple: what started as a system of consonances involving only small whole numbers has turned out to be less simple than at first appeared. For example the major third interval of $\left(\frac{9}{8}\right)^2 = \frac{81}{64}$ and the major sixth $\left(\frac{27}{16}\right)$ and the semitone $\left(\frac{256}{243}\right)$ involve relatively large numbers. However, it

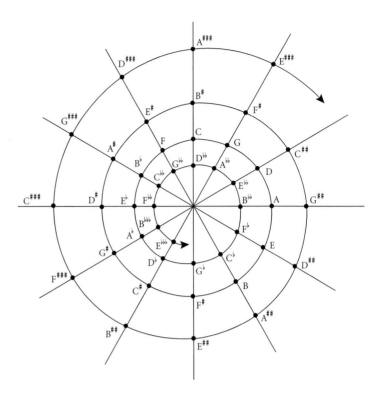

Spiral of Pythagorean fifths.

is important to note that musical intervals until the early renaissance were essentially melodic intervals: they would be perceived as relationships between *successive* notes, rather than as relationships between notes sounded *simultaneously.*

By the time of the early renaissance, polyphonic music had started to develop, and in addition to the harmonic use of octaves, fifths and fourths (hitherto, the only harmonic intervals generally employed), there was a gradual adoption of thirds and sixths. The use of these intervals involved a modification of the Pythagorean tuning under which the third ($\frac{81}{64}$) became slightly flattened to $\frac{80}{64}$, or $\frac{5}{4}$, and the sixth also became slightly flattened, from $\frac{27}{16}$ to $\frac{25}{15}$, or $\frac{5}{3}$.

During the sixteenth century, various attempts were made to modify the Pythagorean scale to incorporate these more consonant thirds and sixths. The most notable of the reformers was Giuseppe Zarlino, choirmaster at St Mark's in Venice. In 1558 he published *Institutioni harmoniche* in which he proposed an alternative mathematical basis for the major scale. He retained the Pythagorean relationships for the octave, fifth and tonic ($4:3:2$), but formalized the earlier *ad hoc* modification of the Pythagorean tuning by adopting the simpler relationships of $6:5:4$ for the perfect fifth, major third and tonic—that is, $\frac{5}{4}$ for the major third and $\frac{6}{5}$ for the minor third. The scale he arrived at, known as the scale of *just intonation*, was as follows:

note	C	D	E	F	G	A	B	C′
frequency	$\frac{1}{1}$	$\frac{9}{8}$	$\frac{5}{4}$	$\frac{4}{3}$	$\frac{3}{2}$	$\frac{5}{3}$	$\frac{15}{8}$	$\frac{2}{1}$
interval		$\frac{9}{8}$	$\frac{10}{9}$	$\frac{16}{15}$	$\frac{9}{8}$	$\frac{10}{9}$	$\frac{9}{8}$	$\frac{16}{15}$

The frequencies of the notes of this scale can all be represented in the form $2^p\cdot3^q\cdot5^r$, where p, q and r are integers, and can be written as follows:

C	D	E	F	G	A	B	C′
1	$3^2/2^3$	$5/2^2$	$2^2/3$	$3/2$	$5/3$	$(3\cdot5)/2^3$	2

We shall refer to this set of notes as **J**. Several new intervals are produced by this scale. For instance, while there are Pythagorean whole tones ($\frac{9}{8}$) for C–D, F–G and A–B, ('major tones'), there are also narrower whole tones ('minor tones') for D–E and G–A of $\frac{10}{9}$. The ratio of these two intervals, $\frac{9}{8}:\frac{10}{9}$, the extent to which they are different tones, is called the *syntonic comma*: $\frac{81}{80} = 3^4/(2^4\cdot5) = 1.0125$, exactly.

The frequency ratios of the just intonation scale occur naturally in the 'harmonic series', and form the basis for playing certain wind instruments. Indeed, in the case of the horn, the technique of playing through using natural harmonics continued until valves were developed during the early nineteenth century. On the natural horn (without valves) the harmonics produce the following written notes.

In this sequence the 2nd, 4th and 8th harmonics correspond to the octave of the scale (that is, they are all the note C), and the 3rd, 6th and 12th harmonics sound G, the perfect fifth. The 9th harmonic sounds the major tone ($\frac{9}{8}$), which is the same in either Pythagorean or just intonation, whereas the 5th and 10th harmonics produce not the Pythagorean major third ($\frac{81}{64}$), but the just major third ($\frac{5}{4}$). Thus far, the natural harmonics are the same as just intonation. However, the 7th/14th, 11th and 13th harmonics (indicated with asterisks) produce notes of $\frac{7}{4}$, $\frac{11}{8}$ and $\frac{13}{8}$, which are wildly out of tune on either Pythagorean or just intonation. Players were expected to coax these notes into tune, the eleventh harmonic being flattened to F ($\frac{4}{3}$) and the seventh harmonic being sharpened up to B♭ ($\frac{16}{9}$). The English composer Benjamin Britten made extraordinary use of these notes in the solo horn prologue of his *Serenade for tenor, horn and strings*, which is scored for natural horn, or for an orchestral horn where the player does not use the valves; the harmonics are indicated in the figure overleaf.

Within a single scale, just intonation formed a reasonably satisfactory solution to problems thrown up by Pythagorean tuning, but the compromise breaks down when one wants to play in another key.

8 8   12   8 8 9 9 6   10 8   8 8 11 11 10 10 9   9 6

10 8   6 6   9   6 6   7   8 6   9 6

Prologue to Britten's *Serenade.*

6 6 7 8 6 9 6 10 6 11   8 8 12   8 8 13 12 8 8 9 9 6   10 4

Transposition with the just intonation scale is even more of a problem than for Pythagorean tuning. When we transpose up by a fifth, we find that the new scale includes *two* new notes: B is transposed to F$^\sharp$, as before, but the D also becomes a new note, an A of $3^3/2^4$, differing by a syntonic comma from the previous A of $\frac{5}{3}$. The reason for this is that the interval G-A in the original scale of C was a 'minor' tone, but became a 'major' tone after transposition.

	C	D	E	F	F$^\sharp$	G	A	B	C'
J	1	$\frac{3^2}{2^3}$	$\frac{5}{2^2}$	$\frac{2^2}{3}$		$\frac{3}{2}$	$\frac{5}{3}$	$\frac{3.5}{2^3}$	2
J^1	1	$\frac{3^2}{2^3}$	$\frac{5}{2^2}$		$\frac{3^2.5}{2^5}$	$\frac{3}{2}$	$\frac{3^3}{2^4}$	$\frac{3.5}{2^3}$	2

On fixed-pitch instruments, such as a harpsichord or organ, this situation made changes of key very difficult. Attempts to overcome the problem meant that alternative keys differing by a syntonic comma had to be provided. One seventeenth-century mathematician who took this issue seriously was Marin Mersenne. In the 31-note keyboard he described and discussed in his *Harmonie universelle* (1636–7), there were no fewer than four keys between F and G!

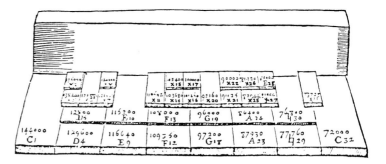

Mersenne's keyboard with 31 notes to the octave.

Two of these (X14 and X15) are G flats differing by a syntonic comma, one for each of the G naturals (again differing by a syntonic comma), one (X16) is an F sharp (for 'F13') and the fourth (X17) is a syntonic comma higher than the F sharp of X16. The following diagram summarizes the relationship of these keys:

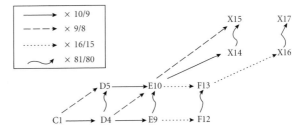

It is interesting to note that such keyboards were actually built: Handel, for example, played a 31-note organ in the Netherlands.

This multiplicity of keys is necessary because successive transpositions of the scale of just intonation generate even more notes between those of the basic set **J** than they did for the set **P**. In this case each transposition produces a new 'black' note, as in the Pythagorean case, but an extra new note is produced, a syntonic comma sharper for upward transpositions and flatter for downward. This arises, as we have seen, because one of the fifth intervals in the just scale is narrow—the interval D–A has ratio $\frac{5}{3} : \frac{9}{8}$ or $\frac{40}{27}$, which is less than $\frac{3}{2}$. In musical terminology, the old submediant is too flat to serve as the new supertonic.

The more transpositions take place, the worse the problems get. The effect of successive upward and downward transpositions of the basic just scale **J** is summarized overleaf.

In practice, modulations into remote keys were not usual at this time (partly, no doubt, for this reason): however, even to use the keys near to C in just intonation required two extra notes per modulation. The systems discussed so far imply infinitely many keys, with the spiral of fifths continuing infinitely, both outwards and inwards: the Pythagorean system **P***, with notes generated by octaves and perfect fifths, and the just system **J***, with notes generated by octaves, perfect fifths and major thirds, both yield infinite sets. So far as the construction of keyboard instruments was concerned, this was not an encouraging state of affairs.

Many attempts were made to develop tuning systems that overcame the difficulties of Zarlino's just system. Amongst these, Francesco Salinas (1530–90) proposed a system called *mean-tone*, in which the two whole tones of Zarlino's system ($\frac{9}{8}$ and $\frac{10}{9}$) were replaced by their geometric mean, thus giving a whole tone interval of $\frac{1}{2}\sqrt{5}$. The interval of the third remained a pure consonance of $\frac{5}{4}$, while the fifth had a ratio of $\sqrt[4]{5}$, which is approximately 1.4953: this is a little less than $\frac{3}{2}$, giving a rather flat fifth. Isaac Newton also spent much time trying to select the best ratios. Believing that seven notes in the octave and seven colours in the spectrum were too much of a coincidence, he even produced a

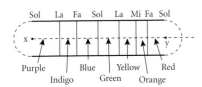

Newton's spectrum scale.

	C	C#	D	D#	E	F	F#	G	G#	A	A#	B	C'	
J^6		$\dfrac{3^7}{2^{11}}$		$\dfrac{(3^5\cdot5)}{2^{10}}$		$\dfrac{(3^7\cdot5)}{2^{13}}$	$\dfrac{3^6}{2^9}$		$\dfrac{3^8}{2^{12}}$		$\dfrac{(3^6\cdot5)}{2^{11}}$	$\dfrac{3^5}{2^7}$		F#
J^5		$\dfrac{3^7}{2^{11}}$		$\dfrac{(3^5\cdot5)}{2^{10}}$	$\dfrac{3^4}{2^6}$		$\dfrac{3^6}{2^9}$		$\dfrac{(3^4\cdot5)}{2^8}$		$\dfrac{(3^6\cdot5)}{2^{11}}$	$\dfrac{3^5}{2^7}$		B
J^4		$\dfrac{(3^3\cdot5)}{2^{10}}$		$\dfrac{(3^5\cdot5)}{2^{10}}$	$\dfrac{3^4}{2^6}$		$\dfrac{3^6}{2^9}$		$\dfrac{(3^4\cdot5)}{2^8}$	$\dfrac{3^3}{2^4}$		$\dfrac{3^5}{2^7}$		E
J^3		$\dfrac{(3^3\cdot5)}{2^7}$	$\dfrac{3^2}{2^3}$		$\dfrac{3^4}{2^6}$		$\dfrac{(3^2\cdot5)}{2^5}$		$\dfrac{(3^4\cdot5)}{2^8}$	$\dfrac{3^3}{2^4}$		$\dfrac{3^5}{2^7}$		A
J^2		$\dfrac{(3^3\cdot5)}{2^7}$	$\dfrac{3^2}{2^3}$		$\dfrac{3^4}{2^6}$		$\dfrac{(3^2\cdot5)}{2^5}$	$\dfrac{3}{2}$		$\dfrac{3^3}{2^4}$		$\dfrac{(3\cdot5)}{2^3}$		D
J^1	1		$\dfrac{3^2}{2^3}$		$\dfrac{5}{2^2}$		$\dfrac{(3^2\cdot5)}{2^5}$	$\dfrac{3}{2}$		$\dfrac{3^3}{2^4}$		$\dfrac{(3\cdot5)}{2^3}$	2	G
J	1		$\dfrac{3^2}{2^3}$		$\dfrac{5}{2^2}$	$\dfrac{2^2}{3}$		$\dfrac{3}{2}$		$\dfrac{5}{3}$		$\dfrac{(3\cdot5)}{2^3}$	2	C
J_1	1		$\dfrac{(2\cdot5)}{3^2}$		$\dfrac{5}{2^2}$	$\dfrac{2^2}{3}$		$\dfrac{3}{2}$		$\dfrac{5}{3}$	$\dfrac{2^3}{3^2}$		2	F
J_2	1		$\dfrac{(2\cdot5)}{3^2}$	$\dfrac{2^5}{3^3}$		$\dfrac{2^2}{3}$		$\dfrac{(2^3\cdot5)}{3^3}$		$\dfrac{5}{3}$	$\dfrac{2^4}{3^2}$		2	B♭
J_3	$\dfrac{(2^4\cdot5)}{3^4}$		$\dfrac{(2\cdot5)}{3^2}$	$\dfrac{2^5}{3^3}$		$\dfrac{2^2}{3}$		$\dfrac{(2^3\cdot5)}{3^3}$	$\dfrac{2^7}{3^4}$		$\dfrac{2^4}{3^2}$		$\dfrac{(2^5\cdot5)}{3^4}$	E♭
J_4	$\dfrac{(2^4\cdot5)}{3^4}$	$\dfrac{2^8}{3^5}$		$\dfrac{2^5}{3^3}$		$\dfrac{(2^6\cdot5)}{3^5}$		$\dfrac{(2^3\cdot5)}{3^3}$	$\dfrac{2^7}{3^4}$		$\dfrac{2^4}{3^2}$		$\dfrac{(2^5\cdot5)}{3^4}$	A♭
J_5	$\dfrac{(2^4\cdot5)}{3^4}$	$\dfrac{2^8}{3^5}$		$\dfrac{2^5}{3^3}$		$\dfrac{(2^6\cdot5)}{3^5}$	$\dfrac{2^{10}}{3^6}$		$\dfrac{2^7}{3^4}$		$\dfrac{(2^8\cdot5)}{3^6}$		$\dfrac{(2^5\cdot5)}{3^4}$	D♭
J_6		$\dfrac{2^8}{3^5}$		$\dfrac{(2^9\cdot5)}{3^7}$		$\dfrac{(2^6\cdot5)}{3^5}$	$\dfrac{2^{10}}{3^6}$		$\dfrac{2^7}{3^4}$		$\dfrac{(2^8\cdot5)}{3^6}$	$\dfrac{2^{12}}{3^7}$		G♭
	C	D♭	D	E♭	E	F	G♭	G	A♭	A	B♭	B	C'	

Just scales.

diagram linking the two; because he wished his scale to be symmetrical, he chose the note D as his starting point, obtaining the following scale:

note	D		E		F		G		A		B		C'		D'
frequency	$\frac{1}{1}$		$\frac{9}{8}$		$\frac{6}{5}$		$\frac{4}{3}$		$\frac{3}{2}$		$\frac{5}{3}$		$\frac{16}{9}$		$\frac{2}{1}$
interval		$\frac{9}{8}$		$\frac{16}{15}$		$\frac{10}{9}$		$\frac{9}{8}$		$\frac{10}{9}$		$\frac{16}{15}$		$\frac{9}{8}$	

Other compromise tunings were also developed, which incorporated some pure consonances: these sounded reasonably satisfactory for keys close to C, but in remote keys they could sound at best unsatisfactory, and at worst excruciating.

Equal temperament

By the beginning of the eighteenth century, it was beginning to be appreciated that for a keyboard to allow unlimited transposition, with no key sounding more in tune than any of the others, it was necessary to divide the octave so that each note was generated by some basic interval: we call this a scale of *equal temperament*. Such ideas had been propounded long before this (in medieval China, for instance). More recently, Galileo Galilei's father Vincenso Galilei had proposed in *Dialogo della musica antica e moderna* (1581) that the scale be constructed from equal semitones with a frequency ratio of $\frac{18}{17}$. It is easy to check that $\left(\frac{18}{17}\right)^{12}$ is about $1.9855\ldots$, a little less than 2, and that $\left(\frac{18}{17}\right)^{7}$ is about $1.4919\ldots$, a little less than $\frac{3}{2}$. Such a scheme would therefore give rather flat octaves and flat fifths, hardly desirable features for the fundamental interval of any scale.

From this proposal it is but a short step to that of Simon Stevin (1548–1620), who suggested making the semitone interval equal to $2^{1/12}$, thereby preserving the octave's frequency ratio of 2. Since $2^{7/12} = 1.4983\ldots$, this choice of semitone still gives slightly flat fifths, but better than those of Vincenso Galilei. $2^{1/12}$ is an irrational number, inexpressible as a fraction p/q and in addition, all of its powers up to the eleventh are also irrational. From a mathematical point of view this is ironical, given that we started out with a criterion for consonance essentially based on the notion of rationality. Of course, $2^{7/12}$ is an extraordinarily *good* approximation to $\frac{3}{2}$, so good that the difference is virtually imperceptible: herein lies the justification for its use. In the following table the frequency ratios for the major scale are compared in Pythagorean, just intonation and equal temperament:

	Pythagorean	just intonation	equal temperament
C	1	1	1
D	1.125	1.125	1.122462...
E	1.265625	1.25	1.259921...
F	1.333333...	1.333333...	1.334839...
G	1.5	1.5	1.498307...
A	1.6875	1.666666...	1.681792...
B	1.8984375	1.875	1.887748...
C'	2	2	2

For ears accustomed to just intonation, the major third of almost 1.26 is noticeably sharp, and thus the extreme consonance of the just major chord $(6:5:4)$ is lost in equal temperament.

Under transposition, we can analyze the behaviour of the equal temperament scale in the same way as we did with the Pythagorean and just scales. The equally tempered major scale has the following notes:

C	D	E	F	G	A	B	C'
1	$2^{2/12}$	$2^{4/12}$	$2^{5/12}$	$2^{7/12}$	$2^{9/12}$	$2^{11/12}$	2

We can again apply the usual transpositions to this set; call it **E**, and let us trace what happens when we arrive at $\mathbf{E}^6$ and $\mathbf{E}_6$. In the following table $2^{1/12}$ is represented by α.

	C	C♯	D	D♯	E	F	F♯	G	G♯	A	A♯	B	C'
$\mathbf{E}^6$		α^1		α^3		α^5	α^6		α^8		α^{10}	α^{11}	
$\mathbf{E}$	1		α^2		α^4	α^5		α^7		α^9		α^{11}	2
$\mathbf{E}_6$		α^1		α^3		α^5	α^6		α^8		α^{10}	α^{11}	
	C	D♭	D	E♭	E	F	G♭	G	A♭	A	B♭	B	C'

The 'new' notes (α^1, α^3, α^6, α^8, α^{10}, α^{11}) now sit symmetrically between the 'old' notes, since they are their geometric means. Hence the transposed sets are identical, so that the keys of G♭ and F♯ are the same. In this way the Pythagorean comma has now been eliminated, and the spiral has been closed into a circle. Six upward and six downward transpositions now give the same set of notes, and we thus arrive at the familiar 'circle of fifths':

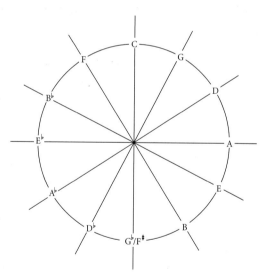

Circle of fifths.

We call the set of notes thus obtained $\mathbf{E}^\star$: the new notes obtained are those generated by $2^{1/12}$, because every note in **E**, $\mathbf{E}^6$ or $\mathbf{E}_6$ is some power of $2^{1/12}$. Moreover, any further transpositions can generate no new notes, so the set $\mathbf{E}^\star$ is a finite set. This is the great advantage of the equal temperament system: there are only twelve notes, and these allow unlimited transposition. The problem of keyboard design is thus solved, because each note now has infinitely many names: the key for F♯ is also that for G♭, as it is also for A♭♭♭ and E♯♯. By the removal of the Pythagorean comma, the spiral has indeed collapsed onto a circle.

The adoption of equal temperament was a lengthy process. Already in the late Elizabethan period (late sixteenth century) there is evidence that English virginal composers (notably John Bull) were modulating so far away from C that a form of equal temperament must have been in use, but as recently as the mid-nineteenth century it was by no means universal, especially in Britain: not one of the British organs at the Great Exhibition of 1851 was equally tempered. However, it is clear that during the early eighteenth century the system was increasingly being exploited. Fischer's *Ariadne musica* (1702), for instance, is a set of miniatures that go through nineteen of the twenty-four major or minor keys. The most famous work to exploit all twenty-four keys is J. S. Bach's *Well-tempered clavier* (1722 and 1738–44). Whether 'well-tempered' meant equally tempered in the modern sense is disputed, but the work includes a prelude and fugue for each major and minor key—hence the usual appellation of 'The 48 preludes and fugues'. Meanwhile a variety of compromise systems co-existed, including for instance the 'Kirnberger III system' which had four just tones, three mean tones, an equal-tempered fifth, nine different semitones and only four major seconds!

The fact that 2^{19} is nearly 3^{12}, and that $2^{7/12}$ is more-or-less $\frac{3}{2}$, is at the root of the equal temperament idea. The question naturally arises as to whether the approximate equation $2^p \approx 3^q$ has any other integer solutions, which might form the basis for an equally tempered scale that gives even better approximations to the just frequency ratios. There are infinitely many solutions, each corresponding to a rational approximation p/q of $\log_2 3$. A good example is $2^{84} \approx 3^{53}$, which leads to $2^{31/53} = 1.49994\ldots$, an excellent approximation to 1.5. This suggests that a structure of 53 notes to the octave (rather than 12) might be better for temperament purposes. In the nineteenth century R. Bosanquet actually made a harmonium with such a subdivision of the octave (see Chapter 5), and the twentieth century saw further exploration of this possibility. Of course, the development of electronic note production in the late twentieth century enabled completely accurate equally tempered systems with any number of notes, as we see in Chapter 9.

The idea of consonance is ultimately grounded in the notion of commensurability, an essential concept in Greek mathematics. We recognise consonance when we perceive a certain number of vibrations of one frequency exactly matching a certain number of another frequency. The Greeks accorded incommensurables a very different ontological status, and it thus remains a powerful irony that irrational numbers should come to the rescue—courtesy of the tolerance of the human ear and cultural conditioning—of the essentially rationally based system that they originally described for constructing a musical scale.

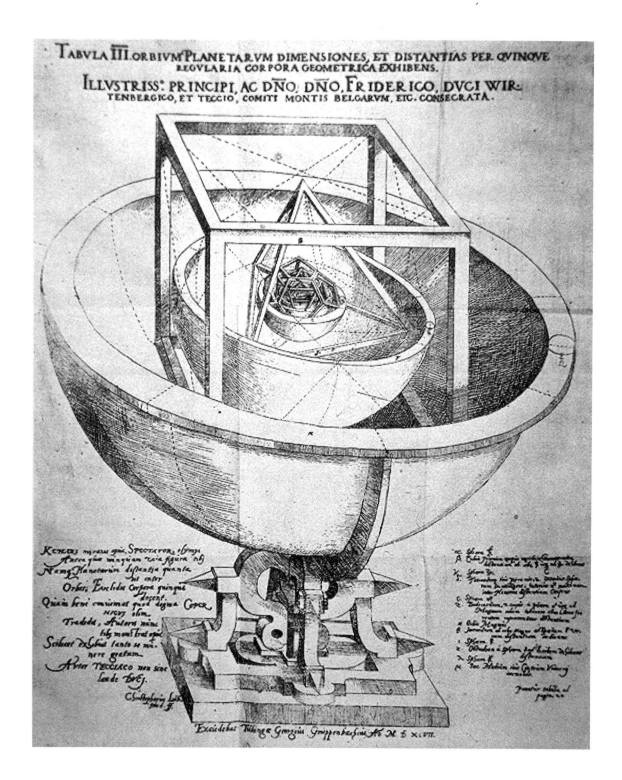

CHAPTER 2

Musical cosmology: Kepler and his readers

J. V. Field

In its more developed form, the mathematical cosmology of Johannes Kepler (1571–1630) presents musical harmony, itself determined by geometry, as a factor in explaining the structure of the Universe. However, his two most influential readers, Marin Mersenne and Athanasius Kircher, recognized the inadequacies of current music theory. Mersenne turns away from the idea of celestial music, but Kircher accepts it, though music itself is perceived not as determined by mathematics but rather as a property built into the Cosmos by its Creator.

In the opening lines of his *Song for Saint Cecilia's Day* John Dryden wrote

> From Harmony, from heavenly Harmony
> This universal frame began: . . .

The poem was first published in 1687—making it an exact contemporary of Isaac Newton's *Mathematical principles of natural philosophy*—but is now probably best remembered in the magnificent musical setting by Handel, which was given its first performance on Saint Cecilia's Day (22 November, New Style) in 1738. By then, and indeed at the time Dryden wrote, the reference to celestial music was no more than a literary device. The main body of the poem is concerned with other matters, but cosmology reappears in the final 'Grand Chorus':

> As from the power of sacred lays
> The spheres began to move,
> And sung the great Creator's praise
> To all the blest above;
>
> So when the last and dreadful hour
> This crumbling pageant shall devour
> The trumpet shall be heard on high,
> The dead shall live, the living die,
> And Music shall untune the sky.

Planetary orbs and regular polyhedra, a fold-out plate from Johannes Kepler's *Mysterium cosmographicum (Tübingen, 1596),* Chapter II; the plate itself carries the date 1597.

As any good Dictionary of Saints will reveal, it is a case of least said soonest mended about the probable connections of the historical Cecilia with music. The connection of music with the origin and structure of the cosmos has a much greater historical credibility. Music theory had been a recognized part of mathematics since Ancient times. Its origins were traced back to the shadowy figure of Pythagoras who, if he was indeed a real person, may have lived in the sixth century BC. Thus, in Ancient, medieval and Renaissance times, to claim that the order of the universe was 'musical' was to claim that it was expressible in terms of mathematics.

We still believe this now. Indeed, mathematical cosmology has proved so powerful that it is perhaps difficult to take a sufficiently cold hard look at the metaphysical basis on which it rests. On the other hand, the explicitly musical cosmologies derived more directly from the Ancient tradition seem sufficiently fantastic to invite instant questioning of their underlying metaphysics—except, of course, in a poetic context such as that provided by Dryden. In his day, those inclined to be unpoetical about cosmology could turn to Isaac Newton for a mathematical explanation of a kind more acceptable in natural philosophy.

Curiously enough, the only natural philosopher to have left a fully worked out mathematical cosmology that uses music theory was the astronomer who supplied the laws from which Newton derived his mathematical theory of gravitation, namely Johannes Kepler. Since Kepler had a high opinion of his cosmological work, it is rather ironic that his own astronomical work did so much to put it out of date. In any case, Kepler saw his cosmological ideas as drawn from an Ancient tradition, essentially from the work of Plato, particularly his dialogue *Timæus*, and from the *Harmonica* of the Alexandrian astronomer Claudius Ptolemy. Ptolemy's treatise is mainly about the theory of music, but it does also contain a sketch of a musical cosmology—geocentric and much simpler than Kepler's fully worked-out heliocentric one. It is, however, the only surviving Ancient text to give a coherent account of what is often called *the music of the spheres*.

Recent research has shown that the complicated combinations of spheres used to explain planetary motion in medieval astronomical texts can in fact be traced back to Ptolemy, and it seems possible that he did believe in solid spheres. Kepler did not. In his first cosmological work, the *Secret of the Universe* (*Mysterium cosmographicum*), he refers to such spheres as 'absurd and monstrous', and he later asks to be shown the shackles that bind the Earth to the sphere that causes its motion. (Since Kepler was a Copernican, he believed that the Earth was one of the planets.) From the point of view of the historian, Kepler is conveniently given to laying his opinions on the line. One is never in doubt, from his first work onwards, that he was a profoundly religious Christian, a totally convinced Copernican, and a devout believer that the Universe is mathematical and to be explained in terms of mathematics. To Kepler, the natural world expresses the nature of its Creator, who is a Geometer, and Man, being made in the image of God, is capable of understanding it in mathematical terms. Indeed, it is his Christian duty to do so.

The first sign of Kepler's interest in music theory is found in connection with cosmology, and it seems likely that this was in fact how it arose. He had presumably become familiar in his youth with the music used in the Lutheran liturgy, but it is difficult to decide what music he might have heard in his later life at the court of the Holy Roman Emperor Rudolf II in Prague. Rudolf had an established taste for all things Italian and italianate, which seems to have extended to music, but the little specific evidence known to historians suggests that while the performers and composers whom he employed were indeed Italian, their music was up to date without being notably *avant garde*. Since in our own time it is precisely the music of the *avant garde*—in particular, that of Claudio Monteverdi—that seems important, one is left with the impression that Kepler may not actually have heard any of the music that, with today's brand of hindsight, can be seen as pointing the way forward.

Kepler's music theory is certainly entirely conventional in its emphasis upon consonance as the sole foundation for music. His earliest references, in the *Secret of the Universe*, are indeed merely to the simple ratios of small whole numbers that define the string lengths corresponding to the standard consonances. Here, Music very clearly takes second place to Geometry, which provides the explicit basis of the cosmological model by which the work is now best known, the system of nested polyhedra and planetary orbs (Kepler defines the orbs as spherical shells that exactly contain the path of the planet). This model is shown at the beginning of this chapter.

None the less, with the characteristic Renaissance faith in the wisdom of the Ancients, Kepler expresses the hope that he will be able to improve his rather clumsy theory of the connection between planetary motion and music once he has read Ptolemy's *Harmonica*. He was apparently unaware that the work was already available in a Latin version by Antonio Gogava, published in Venice in 1562. However, when he did come across this edition, Kepler decided that it was based on a corrupt text—in which today's scholars agree with him. He eventually obtained a Greek manuscript of the work.

Kepler had hoped that his astronomical calculations of more accurate planetary orbits, using the observations made by Tycho Brahe, would confirm the correctness of the polyhedral cosmology described in his *Secret of the Universe*. In the event, the new more accurate orbits did not agree more closely with the theory, which Kepler had in any case already begun to modify. The result was his *Five books of the harmony of the world* (*Harmonices mundi libri V*), published in Linz in 1619.

In this work, everything starts from geometry. The first two books are concerned to establish hierarchies of regular polygons, the rank of a

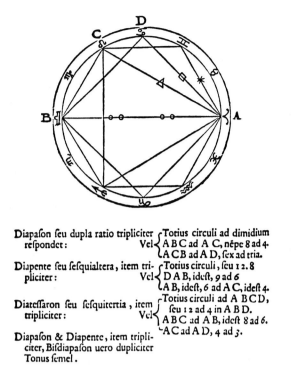

Diapafon feu dupla ratio tripliciter refpondet: Vel {Totius circuli ad dimidium A B C ad A C, nêpe 8 ad 4. A C B ad A D, fex ad tria.

Diapente feu fefquialtera, item tripliciter: Vel {Totius circuli, feu 12. 8 D A B, ideft, 9 ad 6 A B, ideft, 6 ad A C, ideft 4.

Diateffaron feu fefquitertia, item tripliciter: Vel {Totius circuli ad A B C D, feu 12 ad 4 in A B D. A B C ad A B, ideft 8 ad 6. A C ad A D, 4 ad 3.

Diapafon & Diapente, item tripliciter, Bifdiapafon uero dupliciter Tonus femel.

Figure 1.   Musical consonances and astrological 'aspects' (angles between heavenly bodies that were believed to modify their degree of influence), from Claudius Ptolemy, *Harmonica*, translated by Antonio Gogava, Venice, 1592, p. 144.

particular figure being established, in the first book, by how many operations—each using only straightedge and compasses—are required to inscribe its side in a circle of given radius. In the second book, rank is established by the polygon's capacity to participate in forming tiling patterns or polyhedra. There are numerous echoes of Plato's *Timæus* and Ptolemy's *Harmonica*.

However, Kepler has departed from Ptolemy's simple linking of consonances with astrological 'aspects' (see Figure 1). It is the ranking of polygons in the second book that determines their importance in astrology (as explained in Book IV), whereas in the third book the ranking of the first book is applied to music theory, the higher-rank polygons dividing the circle (in the manner shown in Figure 1) to give the more perfect consonances among the ratios of the parts. The result is a ranking of consonances that corresponds to that given by Gioseffo Zarlino in his *Institutions of harmony* (*Istitutioni harmoniche*, Venice, 1558), where they are derived from pure numbers. However, Kepler correctly recognizes that this system corresponds with that of Ptolemy, so Ptolemy gets the credit for it and the name of Zarlino is mentioned only once. Having devised his own geometry-based version of the system, Kepler proceeded, in his fifth book, to apply it to his—that is God's—heliocentric planetary system.

Harmonia bimorum		Apparentes diurni diurni. Prim Sec.	Harmonia singulorum propria Prim. Sec.	
Diver. a 1	Conf. b 1	♄ Aphelius 1.46.a. Perihelius 2.15.b.	Inter 1.48 & 2.15,	est $\frac{4}{5}$ Tertia major.
d 3 c 8	c.2 d 1	♃ Aphelius 4.30.c. Perihelius 5.30.d.	Inter 4.35. & 5.30.	est $\frac{5}{6}$ Tertia minor.
f 1 e 5	e 5 f 2	♂ Aphelius 26.14.e. Perihelius 3.81.f.	Inter 25.21. & 38.1.	est $\frac{2}{3}$ Diapente
h 12 g 3	g 3 h 5	Tel.Aphelius 57.3.g Perihelius 61.18.h	Inter 57.28. & 61.18.	est $\frac{15}{16}$ Semitoni:
k 5 i 1	i 3 k 3	♀ Aphelius 94.50.i. Perihelius 97.37.k.	Inter 94.50. & 98.47.	est $\frac{24}{25}$ Diesis
m 4	l 5	☿ Aphelius 164.0.l. Perihelius 384.0.m.	Inter 164. 0. & 394. 0:	est $\frac{5}{12}$ Diapason. cum tertia minore

Figure 2. A table of daily planetary motions in arc, seen from the sun, from Johannes Kepler, *Harmonices mundi, libri V*, Linz, 1619, Book V, Chapter IV.

Kepler looked for, and failed to find, musical proportions in various quantities in the Solar system, for instance, in the periods of the planets—a case in which his lack of success is displayed in the form of two tables. He found the ratios he was looking for in the extreme speeds of the planets, that is their speeds when they are nearest to the Sun (perihelion) and furthest from it (aphelion). He uses both the ratios of extreme speeds for each individual planet and the ratios of extreme speeds of pairs of neighbouring planets. The results are displayed in a table, which is reproduced in our Figure 2.

The speeds are expressed as motions in arc, that is angular motions as seen from the Sun—which means that to a Copernican they represent actual motions in space (the Sun being assumed at rest). The second column (headed *Apparentes diurni*, that is apparent daily motions) gives the angular motions, in minutes and seconds of arc, starting with the extreme values for Saturn and then working inwards through the planets of the system. Each speed is given a letter of the alphabet, to identify it when it is used elsewhere in the table. The third column gives the ratios obtained for individual planets, and notes their correspondences with particular musical intervals. The first column gives the ratios obtained from extreme speeds of neighbouring pairs, starting with Saturn and Jupiter, the first ratio being the 'diverse' ratio, of the aphelion (minimum) speed of Saturn to the perihelion (maximum) speed of Jupiter, and the second being the 'converse' ratio, that of the maximum speed of Saturn to the minimum speed of Jupiter.

From these ratios, that is from the intervals they define, and setting an arbitrary note as starting point, Kepler constructs two musical scales, shown in our Figure 3, one in *cantus durus* (which does not quite correspond to the modern diatonic major scale) and one in *cantus mollis* (which

Figure 3.   Scales in *cantus durus* and *cantus mollis*, from *Harmonices mundi*, libri V, Linz, 1619, Book V, Chapter V. The symbol resembling the modern sign for a double sharp represents a sharp; that resembling the modern sign for a flat represents a flat.

does not quite correspond to the modern diatonic minor scale). Kepler has written 'free' (*Vacat*) beside one note in each scale. These pitches are covered later, when he gives a complete compass of notes for each planet (see Figure 4).

In the scales shown in Figure 3, some notes are marked *fere*, which means 'approximately'. None the less, closer inspection shows that the approximations are very good ones, by any reasonable standards. Unlikely as it may seem, the numerical relationships which Kepler has found, and which he has chosen to express in musical form, are in very good agreement with the values deduced from Tycho's observations.

In fact, twentieth-century observational values of the velocities concerned also yield 'musical' ratios (the modern definition of a musical ratio being, for astronomical purposes, that it involves small whole numbers, 'small' being deemed to go up to about 7). This in effect merely confirms what we in any case know: that Kepler's planetary orbits are in good agreement with modern ones, the agreement being so good, in fact, that in assessing it one has to allow for long-term (secular) changes since the late sixteenth century (when Tycho's observations were made). The real puzzle, for today's experts on celestial mechanics, is how the Solar system came by these ratios, which are now usually called *resonances*. A particularly spectacular set are observed among the periods of the moons of Jupiter.

Kepler's astronomy was, as we should say today, state of the art. So too, but much less satisfactorily, was his music theory. At this time, theoreticians of music simply could not cope with the way in which composers made increasing and systematic use of dissonance for expressive or dramatic purposes. Nor had adequate theoretical solutions been found to the problems associated with tuning a number of different instruments to play different lines of music together.

Figure 4. Compasses of the planets, from Kepler, *Harmonices mundi, libri V, Linz, 1619, Book V, Chapter VI.*

Performers worked with their own practical approximate methods, but music was still claimed as one of the mathematical sciences. Intellectual coherence between theory and practice was not to be achieved until the eighteenth century, in the musical treatises of Jean-Philippe Rameau.

Kepler tried fairly hard to prevent his *Harmony of the world* being put on the Index of Prohibited Books, even pointing out that his system of harmonies would be equally real in the Tychonic model of the planetary system, in which all bodies except the Earth moved round the Sun, which itself moved round the Earth, carrying them with it. This protestation did not work. Indeed, the 1630 Index for Rome simply bans all Kepler's books. So one might not expect to find loyal supporters of Papal authority among Kepler's readers—and particularly not after the condemnation of Galileo in 1633. But one would be mistaken. It is, of course, impossible to know how many people actually read the *Harmony of the world*. One can only consider those who chose to mention the work in print. They were few. However, in the forty years following the publication of Kepler's work there appeared two extremely bulky treatises whose titles immediately suggest comparison with it, namely the *Harmonie universelle* of Marin Mersenne, published in Paris in 1636, and the *Musurgia universalis* of Athanasius Kircher, published in Rome in 1650.

Mersenne's family were peasants. He was born in a small village near the town of Le Mans, and his rather old-fashioned crabbed handwriting is probably a relic of his education at the village school. He joined the religious order known as the Minims. His working life was spent in Paris, where he became the centre of what amounted to an informal academy whose members exchanged news and information about mathematics and natural philosophy. Its members and correspondents included many distinguished mathematicians, such as Girard Desargues, René Descartes and the young Blaise Pascal, who was introduced by his father Etienne Pascal. That is, as a clearing house for scientific information, Mersenne was extremely impressive.

It is rather more difficult to be sure about his personal intellectual qualities. Most of his works are extremely long—upwards of two thousand pages, folio—and they sometimes seem to be putting forward two contradictory views almost simultaneously. For example, in a little book published as an introduction to his huge treatise on music, he takes about fifteen pages (octavo) to give a detailed account of the horoscope for a perfect musician, but on the final page devotes a few lines to saying horoscopes are all nonsense. The historian may be grateful for the information that the exemplar of the perfect musician is Orlande de Lassus (*c*.1532–94)—though it should be pointed out that astrologers regularly falsified dates in order to obtain more appropriate horoscopes, so the date Mersenne gives is no help in guessing Lassus' actual date of birth—but the blood surely runs cold as one contemplates the possibility of similarly brief (and thus easily overlooked) contradictions of wordily expressed opinions in the two thousand pages of the *Harmonie universelle*. In the circumstances, it is perhaps lucky that the treatise turns out to have surprisingly little to do with our present concerns. What is universal about the work is not its concern with the cosmos but rather its total scope. Celestial music is dealt with rather briefly.

It is clear that Mersenne not only understood and enjoyed music, but was also very knowledgeable about the music of his day. His work is, in fact, important for the number of musical pieces it preserves. It also contains a huge array of illustrations of musical instruments. Reference to the *table des matières* (an index) yields several references to 'Galileo examined' but none explicitly to Kepler, although his name does occur within some entries, for instance in 'Kepler's octave divided into twelve parts' (under D, because it reads 'Diapason de Kepler . . .'). Either Protestantism was even hotter to handle than being 'vehemently suspected of heresy' by the Inquisition (its sentence on Galileo) or Mersenne simply was not interested in the musical cosmology of Kepler's *Harmony of the world*. He certainly had read the work, since he makes several references to it, for instance in Proposition X of his book 'On the usefulness of harmony' (the penultimate item in the treatise, separately paginated). Here there is an illustration—taken from Robert Fludd's *An account of both worlds . . .* (*Utriusque cosmi . . . historia*, Oppenheim, 2 vols., 1617, 1618), without acknowledgement—to show very simple consonances among the spheres of a Ptolemaic planetary system. When Kepler objected that no astronomer could believe in this system any more and that Fludd was concerned with a world of his own imagining, Fludd retorted that his harmonies were in the Soul of the World, whereas Kepler's were merely in its Body. So much for agreement with observation! Mersenne, who is usually seen as a proponent of the new natural philosophy, is here apparently putting forward something completely out of date. He has earlier briefly dismissed Copernicanism, and referred to Kepler's ideas as fanciful, so perhaps we should see his parading this ostentatiously old-fashioned cosmology as a defence against a possible imputation of advocating innovation in astronomy. One could, also, put Mersenne back on the side of the angels by suggesting that he did not have much time for such old-fashioned stuff as celestial music.

Though one may wonder about the astronomy, the music theory in Mersenne's treatise is certainly intended to be taken seriously. Indeed, Mersenne even attempts to explain the emotive effects of music, with reference to composers' use of dissonances. There was Ancient authority for the belief that music could cause emotions, specific kinds of music exciting specific emotions. The standard exemplar of this power in action was Orpheus, who (according to the song in Shakespeare's *Henry VIII*, of 1612)

> . . . with his lute made trees
> And the mountain tops that freeze,
> Bow themselves when he did sing . . . .

Orpheus, 'taken from an Ancient marble', duly appears in the frontispiece to Mersenne's treatise, although the quotation underneath is from the Psalms (see Figure 5 overleaf).

Mersenne's explanations of the 'effects' of music are not very satisfactory. They deal mainly with the rather gentle use of dissonance found in contemporary French music—which was, of course, what Mersenne knew. Like Kepler, Mersenne is essentially dealing with the kind of dissonances habitually used in the later work of Lassus. Kepler was almost certainly relying on a secondary source that specifically referred to Lassus, and Lassus' reputation was high in France at the time Mersenne wrote. In any case, in contrast to the mathematical nature of the rest of his exposition, Mersenne's treatment of dissonances and their 'effects' is qualitative and imprecise, and does not, it seems, cause him to re-think the otherwise conventional theory of music that he puts forward. However, one can see that a genuine love of the music of his own time was having an effect. The same is true of the music theory of Athanasius Kircher, but in his case the effect seems to have been much more drastic.

Athanasius Kircher was a Jesuit. He rose to high rank in the foremost Jesuit College, the one in Rome. Like Mersenne he wrote a great deal, and often at great length. His intellectual interests were multifarious, and various passages from his works are usually cited as marking the beginning of modern interest in, say, deciphering Egyptian hieroglyphics, considering the origins of volcanoes, or setting up public museums. In his own day Kircher was certainly taken very seriously, but in ours short extracts, punctuated with some of his many elegant illustrations, can easily give the impression that he was, to put it bluntly, a weirdo. As we shall see, the final book of his treatise on music tends to lend weight to this impression.

However, the work as a whole does not. Kircher's *Musurgia universalis* is, like Mersenne's *Harmonie universelle*, a musical encyclopaedia, designed to teach the reader about music, but it is also about the place of music in Creation. Even more than Mersenne, Kircher quotes

HARMONIE
VNIVERSELLE

Ex antiquo marmore Illustrissimi Marchionis Mathei Romæ.   N. le Bas fecit

Nam & ego confitebor tibi in vafis pfalmi veritaté tuam:
Deus pfallam tibi in Cithara, fanctus Ifrael.   *Pfalme 70.*

Figure 5. Frontispiece to Mersenne's *Harmonie universelle*, Paris, 1636, showing Orpheus, 'from an ancient marble in the collection of the most illustrious Marchese Mattei, Rome'. Mersenne's numbering of the Psalm is taken from the Vulgate; in the Authorised Version, the passage is *Psalm 71*, verse 22: 'I will also praise thee with the psaltery, *even* thy truth, O my God: unto thee will I sing with the harp, O thou Holy One of Israel.'

complete musical compositions, usually absolutely modern, and with the highest praise for their excellence. One of the interesting things about these judgements is that several of the composers Kircher praises were also admired, about fifty years later, by Johann Sebastian Bach. Moreover, when it comes to dealing with the 'effects' of music, Kircher cites recent (and not so recent) Italian examples, including, for 'pain' (*dolor*), a passage from a madrigal by Carlo Gesualdo, a composer whose music was famous for its dissonance. In fact, Kircher cites Gesualdo several times, and also praises Monteverdi as the composer most skilful in portraying emotion. Both compliments are the more weighty when we remember that it was generally taken for granted that newer music was better than older music. All in all, Kircher seems prepared to accept a considerable degree of dissonance as a natural component of music. But he cannot construct a mathematical theory that explains how this works, and the tuning problem is not mentioned.

It seems likely that an awareness of these shortcomings in the mathematical theory played a part in determining the overall structure of *Musurgia universalis*. Kircher begins with anatomy, that is, with the anatomy of the parts of animals that make sounds (see Figure 6).

One animal is singled out for a chapter all to itself: the three-toed sloth (habitat South America). Kircher's informant is said to be a missionary by the name of Johannes Torus. The sloth is alleged to sing up and then down a perfect musical scale (an 'ordinary scale', Kircher says, thereby ducking a few questions). Now this, as one might guess, turns out not to be quite as it seems. The singer is not the sloth but the Common Potoo, which is a bird. The potoo usually sings at twilight, flattened against its perch on a tree beside the river. The sounds are eerily pure, and do indeed form a descending diatonic scale. The belief that the sound is made by the sloth is a piece of local folklore: to this day, local Indians, who know a great deal about the animals they hunt, know little about the sloth and find it rather an eerie animal. The National Sound Archive, who kindly told me about the potoo, and played me the appropriate tapes, did something of a double-take on discovering that the folklore story went back to the seventeenth century. Actually, it goes back at least to the sixteenth. After the sloth, we have chapters about bird song, complete with songs written out on staves (see Figure 7).

After this, Kircher turns to mathematical music theory, with, as we have seen, not completely satisfactory results. Like Mersenne, he also considers musical instruments, and explains, for instance, which note corresponds to each string of a particular kind of viol. He devotes a whole book to the 'effects' of music and another one to anecdotal evidence of its magical powers.

Undaunted by the inadequacy of his music theory to explain human ('artificial') music in mathematical terms, in his final book (Book X)

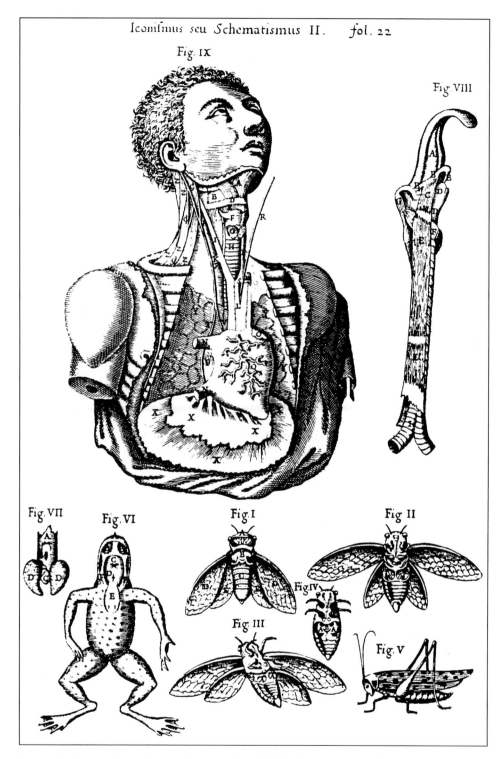

Figure 6.   Anatomy of the organs that produce sound in animals, from A. Kircher's *Musurgia universalis*, Rome, 1650, Plate II, opposite page 22.

Figure 7. Bird songs from A. Kircher, *Musurgia universalis*, Rome, 1650, Plate III, opposite page 30. The long song at the top is that of the nightingale. Note that the parrot is saying 'hello', in classical Greek—Kircher had been a Professor of Greek.

Iconismus XXIII. Harmonia primi diei fol. 366 Tomo 2.

HARMONIA NAS—CENTIS MVNDI

Harmonia II.diei

Harmonia III.diei

Harmonia V.diei

Harmonia IV diei

Harmonia VI.diei

Figure 8.   Athanasius Kircher, *Musurgia universalis*, Rome, 1650, Plate XXIII. 'Harmony of the World at its birth', Tomus II, p. 266, beginning of Book X, illustration of the Organ of the World. The planetary system shown at the left is geocentric.

Kircher goes into some detail on the subject of celestial music, which he links not only with the structure of the cosmos but also (like Dryden) with its origin. There is, however, nothing mathematically precise about this, as can be seen from the illustration of the Organ of the World, shown in Figure 8.

One could clearly decide to take this on a high metaphorical level. However, Kircher does not leave the matter there. He proceeds to discuss harmonies among the motions of the planets. He dismisses Kepler, by name, as misguided, and then, without attribution, uses all Kepler's numbers. There is no explicit dismissal of Copernicanism. In fact, Kircher simply reprints the table we showed in our Figure 2, and more than a page of the accompanying text is taken *verbatim* from the *Harmony of the World* (which, being on the Index, was in principle unavailable to Kircher's readers). Kircher was, of course, quite right to guess that Kepler's numbers were the best then available, but his method of using them leaves one wondering what is the Latin for *chutzpah*.

Kircher does actually state that music has the force of a mathematical organizing principle in Creation, but he does not go into details. There is, however, a certain amount of rather crude astrology. The work in fact ends in a decidedly messy way. However, it is not only in that respect that we are far from the orderly mathematical cosmology put forward by Kepler. If one looks at the structure of the works as a whole, it seems that Kircher has recognized the inadequacy of the mathematical theory of music, and has instead devised a theory that makes music itself a natural property of the World, as created by God. Human music has its counterpart, or even, he hints, its origin in the sounds made by animals and birds. There is a suggestion that South America may have been the site of the Garden of Eden (hence the song of the sloth). That is, Kircher is involving the world of living things in what was traditionally an area for abstract mathematical theorising. The chief thrust of the Scientific Revolution of the seventeenth century was to extend the domain of mathematics (it is no accident that Newton called his work *Mathematical Principles of Natural Philosophy*), so Kircher was certainly out on a limb in this respect. However, his wealth of information about the natural world, and about music, fits in rather well with another characteristic of the Revolution: the systematic accumulation of observations.

Kepler's work is important as the first mathematical cosmology (however bizarre it may look to his present-day heirs), and it is in good agreement with astronomical observation. It is, however, in rather bad—if, by then, standardly bad—agreement with musical observation and practice. A recognition of this failure of the standard music theory may account for Mersenne's apparent unwillingness to take cosmic music seriously. However, Mersenne, in France in the 1630s, did not have to face the inadequacy of music theory to explain practice in quite such a radical manner as Kircher did in Rome in the 1640s. Kircher is, moreover, notable for having done nothing to try to make the problem look less intractable. Instead he chose to go round another way, making music seem less specially mathematical in the process. This did not actually help very much. The solution was eventually to be found in more complicated mathematics, and, as in so many other branches of natural philosophy, by ruling out certain questions as beyond the range of a reasonable theory. The problems that got thrown out here were those of giving precise accounts of the 'effects' of music and of the musical significance of dissonance. Moreover, within about a century of Newton's work, it was clear to astronomers that one could no longer equate the Solar system with the Cosmos. Thus, what was cosmology to Kepler, Mersenne and Kircher became, for Newton's successors, no more than a theory of the Solar system. All the same, since human ego-centricity gives human thought a persistent tendency to geocentricity, 'the music of the spheres' seems destined to remain a part of poetic vocabulary.

PART II

The mathematics of musical sound

CHAPTER 3

The science of musical sound

Charles Taylor

This chapter complements the others by describing practical demonstrations and experiments. In recent years a good deal has been said about the differences in experiments in an elementary physics laboratory, mathematical theory, and real musical instruments. In fact there are no real differences except those arising from too simplistic an approach.

Sound of any kind involves changes of pressure in the air around us; for example, in ordinary speech the pressure just outside the mouth increases and decreases by not more than a few parts in a million. But to be detected by our ears and brains as sound, these changes have to be made fairly rapidly. This can be demonstrated easily by inflating a balloon and then gently squeezing it between thumb and finger. This creates quite large pressure changes without any attendant sound; but inserting a pin creates a change that can very readily be heard.

Scientists study the nature of the pressure changes using a cathode-ray oscillograph that draws a graph of the pressure as a function of time. It is interesting to look at the traces corresponding to a wide variety of sounds and try to relate what is perceived by the ear–brain system with what is simultaneously perceived by the eye–brain system. It proves to be impossible to make any but the broadest generalizations about a sound by observing only its oscillograph trace. As an example, it is not easy to differentiate between the oscillograph traces of the end of the first movement of Mendelssohn's *Violin concerto*, a symphony orchestra 'tuning up', and the chatter of an audience waiting for a lecture to begin (see overleaf), although aurally they are completely different.

One of the most astonishing properties of the human brain is that of recognizing sounds in a split second. For example, if, a dozen subjects are all asked to repeat the same word, an audience has no difficulty in understanding what is being said. But for each one of the twelve, the corresponding oscillograph traces is completely different and it is virtually impossible to find common features.

So here we have two different ways of presenting the same information: the brain has little problem in interpreting the aural form, but the visual form presents far greater difficulties.

Charles Taylor demonstrating an oboe to some children.

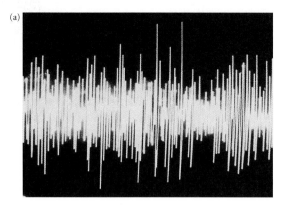

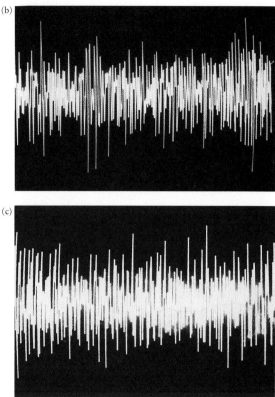

Oscillograph traces of three different sounds:
(a) the end of the first movement of the Mendelssohn violin concerto;
(b) a symphony orchestra tuning up;
(c) an audience waiting for a lecture to begin.

The part played by the brain

Having introduced the topic of aural perception, we next elaborate on its remarkable features, since it affects practically every experiment done in the field of musical acoustics.

We first notice that the pressure of the air can only have a single value at a particular point at a particular time. So, if you listen to a large orchestra of seventy players, each instrument creates its own characteristic changes in pressure, but they all add together to produce a single sequence of changes at the ear and there is only one graph of pressure against time that represents the sum of the changes produced by all the instruments. Yet, with surprisingly little effort, a member of the audience can listen at will to the different instruments. The problem of disentangling these instrumental components from the single graph would be extraordinarily difficult for a computer, unless it were given all kinds of clues about the nature of each different instrument, but the human ear–brain system performs the miracle in a fraction of a second.

One of the factors that makes this possible is the learning ability of the brain. Stored in our brains we all have the characteristic features of all the various instruments that we have heard before and these can be drawn on subconsciously to aid the disentangling process.

An interesting example of this learning process is as follows. A recording of synthetic speech can be created by first imitating the raw sound of the vocal chords by means of an interrupted buzz on one note, adding chopped white noise to represent ss, sh and ch sounds, and then introducing just one formant for each vowel. An audience is unable to recognize the sentence that has been synthesized. However, having been told what the sentence was, they have no difficulty in recognizing it on a second hearing.

This ability of the brain, both to memorize sounds, and to identify similar sounds in the memory banks at great speed, is vital to our existence, but is also a great nuisance in psycho-acoustic research. Its importance lies in the way that we can rapidly identify sounds that indicate danger, in the way that we learn to speak as babies, and in the way that we can adapt to very distorted sounds and in many other activities. Adaptation to distorted sounds is illustrated if one listens to messages being relayed over 'walkie-talkie' systems to the police, to pilots in flying displays, and in other circumstances where those used to the system have no difficulty in understanding the messages, but outsiders find the speech hard to follow.

The problem in psycho-acoustic research arises because the very act of performing the first experiment produces changes in the memory banks of the subject. For example, consider an experiment on pitch perception where a participant is asked to compare groups of sounds and to say which is the highest in pitch. Once the first group of sounds has been heard it is impossible for the subject to ignore those sounds, and the response, at a latter stage, even to the same group of sounds, is very rarely the same.

Another of the many remarkable properties of the brain that plays a part in our aural perception is that of ignoring sounds which are of no importance to us. If a series of sounds—such as a baby crying, a dog barking or a fire engine's siren—were played while someone continued to speak, then the listeners will continue to hear what is being said, because they rapidly identify the extra sounds as of no personal relevance.

Differences between music and noise

The above examples are of relatively complex sounds and it is difficult to draw clear scientific distinctions between music and noise with sounds of this complexity. The two simplest kinds of sounds that occur in studies of sound are white noise and a pure tone. Musically useful sounds consist of mixtures and modifications of these two basic kinds of sound—pure tones and noise.

The oscillograph trace of white noise shows no element of regularity at all. The only variable parameter is that of the amplitude, which

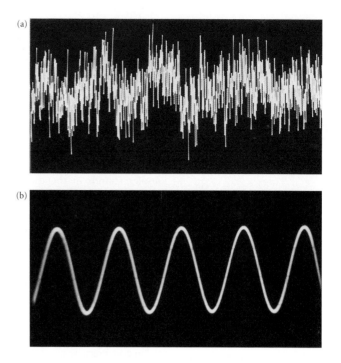

(a)

(b)

(*a*) oscillograph trace for white noise
(*b*) oscillograph trace for a pure tone

corresponds to the loudness of the sound. (It is possible, of course, to 'colour' noise by filtering out various frequency components, but then it can no longer be described as 'white'.)

The oscillograph trace for a pure tone is that of a sine wave and is completely regular. There are now two parameters that matter: the amplitude which, as before, relates to the loudness, and the frequency which relates to the pitch of the note. Many textbooks tend to keep these two parameters separate, but in fact they are linked, again because of the mechanism of perception. For example, listen to a pure tone of frequency 440 Hz (the note with which the tuning of orchestral instruments is checked) at a relatively low amplitude. Then, without changing the frequency, increase the amplitude very suddenly. The loudness will increase, and many people will also detect a change in pitch. With a fairly large audience one usually finds that about a half hear the pitch go down, rather fewer hear it go up, and a few hear no change. This is a dynamic effect that only occurs with sudden changes and only with fairly pure tones.

Sources of musical sounds

Many common objects have a natural frequency of vibration that can be excited by striking or blowing. All kinds of tubes, or vessels with a narrow opening, for example, will emit a sound if the opening is struck with the flat of the hand. In this case it is the air that is vibrating and, if

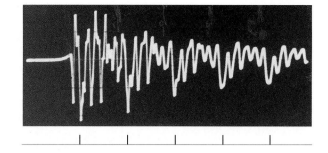

Oscillograph trace from a tube when a cork in one end is suddenly withdrawn; the vertical lines correspond to the natural resonant frequency of the tube.

the natural frequency lies within the sensitivity range of the human ear (about 30–18000 Hz in young people, although the upper sensitivity declines rapidly with age), a musical sound is heard. When a cork is suddenly withdrawn from the end of a tube a compression wave travels back and forth in the air in the tube. Although its amplitude rapidly decays, as shown above, the time taken for each transit determines a discernible musical pitch in the short-lived sound.

In order to convert this into a usable musical instrument, we must feed in energy to keep the wave travelling up and down for as long as the note is required. This can be done by blowing across the end with such a speed that the edge tone generated as the air jet strikes the edge has an oscillatory frequency that matches the natural frequency of the tube. Alternatively a reed can be used. All reeds are, in effect, taps that allow pulses of air to pass through at a well-defined frequency, which again can be made to match that of the pipe. The lips form the reeds in the brass family of instruments, single or double strips of cane form the reeds of the woodwinds.

Similar arguments can be applied to the vibration of strings. Transverse vibration of strings can be excited by striking (as in the piano or clavichord), or by plucking (as in the harp, guitar or harpsichord). But to convert such short-lived notes into those of much longer duration, energy must be fed in to maintain the vibration. In modern electric guitars various forms of electronic feedback can be used, but the traditional method, used in the orchestral string family, is by bowing. This depends on the difference between the static and dynamic frictional properties of resin. Powdered resin adheres to the microscopic scales of the horse-hair used in bows: when the bow is placed on the string the static friction is high but when moved to one side the string sticks to it and is also moved to the side. Eventually the restoring forces created in the string overcome the static friction, and the string starts to slip back to its neutral position. Dynamic friction, which is very much lower than the static friction, allows the string to move easily under the bow, overshoot the neutral position, come to rest, and then be picked up once more by the static frictional force to repeat the cycle.

Harmonics, overtones, and privileged frequencies

Although most objects have a natural vibration frequency, the real situation is much more complicated. An easy way to approach these complications is to think of a child's swing. The oscillation can be kept going by giving a slight push once in every cycle of the swing—but the timing is all important and it is just as easy to bring the swing to a standstill if the push is applied at the wrong moment. The right moment is just after the swing has started to accelerate from one of the extreme positions and the push must obviously be in the same direction as the movement of the swing. But the swing can also be kept going if a push is given every second time the swing reaches the optimum position, or every third time, and so on. Equally, if the person pushing gives a push (some of which, of course, will not connect with the swing) at twice the natural frequency of the swing, or at three times the natural frequency, the pushes that connect with the swing will still be at the right frequency to maintain the oscillation.

Consider the tube discussed earlier. The oscillation can be maintained if the hand is repeatedly slapped on the end of the tube at its natural frequency f. But, as with the swing, it could equally well be excited at frequencies $2f$, $3f$, $4f$, $5f$, etc. and also at frequencies $\frac{1}{2}f$, $\frac{1}{3}f$, $\frac{1}{4}f$, etc. Indeed, it can also be excited at $\frac{2}{3}f$, $\frac{3}{4}f$, $\frac{5}{3}f$, and at many other possible frequencies.

The frequencies commonly discussed in connection with musical instruments are f, $2f$, $3f$, $4f$, etc., which are usually termed *harmonics* (see Chapter 1). In practice, because of end effects, the effect of the diameter of a pipe, and many other complications, a simple tube will not resonate precisely at the harmonic frequencies—but in spite of this musicians still tend to call them harmonics. Scientists know them as *overtones*.

The remaining frequencies of the type $\frac{1}{2}f$, $\frac{1}{3}f$, or $\frac{3}{2}f$, $\frac{5}{3}f$, etc., are known as *privileged frequencies* (and strictly speaking, the harmonics are privileged frequencies as well). The following table shows a list of the harmonics and privileged frequencies for a tube open at both ends with a basic natural frequency of 240 Hz: the numbers in bold type are the true harmonics.

120	**240**	360	**480**	600	**720**	840	**960**	1080	**1200**
60	120	180	**240**	300	360	420	**480**	540	600
40	80	120	160	200	**240**	280	320	360	400
30	60	90	120	150	180	210	**240**	270	300
24	48	72	96	120	144	168	192	216	**240**

Notice that some of the privileged frequencies (such as 120 and 60) occur more than once, and if the table were still further extended others would occur. These frequencies are easier to excite than the ones that occur only once.

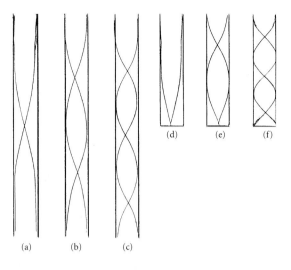

Traditional diagrams showing the graphs of the displacements in open and closed pipes:
(*a*) open, frequency *f*
(*b*) open, frequency *2f*
(*c*) open, frequency *3f*
(*d*) closed, frequency *f*
(*e*) closed, frequency *3f*
(*f*) closed, frequency *5f*.

Impedance view of the behaviour of tubes

In the elementary approach to the behaviour of vibrations in tubes, use is made of the fact that a compression becomes an expansion on reflection at an open end, but stays a compression when reflected from the end of a closed pipe (see above). This must obviously be so, as the reflection at the end of an open tube must add to the outgoing wave to produce no excess pressure, and must therefore be an expansion. For the closed pipe there is obviously maximum pressure at the end.

Problems arise if the pipe is not precisely cylindrical for the whole of its length and it is then no longer possible to draw convincing diagrams based on the simple theory. Measurement of the input impedance of a tube as a function of frequency leads to a more satisfactory argument.

The figure (*a*) overleaf shows such a diagram based on the work of Arthur Benade. The difference in behaviour between edge-tone instruments and reeds can be explained without assumptions about open or closed ends. Edge tone excitation involves only small changes in pressure, although the displacements are high. Thus it is a low impedance device (analogous to a low-voltage high-current electrical device) and, as can be seen from the diagram, this leads to a full series of harmonics. A reed, on the other hand, involves relatively low air flow but high pressure changes, and is thus a high impedance device, which can be seen from the diagram to involve only the odd harmonics, but the fundamental is an octave lower than that for a low impedance instrument. The input impedance curve for a pipe with a series of side holes (as in most woodwind instruments) is shown in figure (*b*); the existence of a cut-off frequency can be clearly seen. The input impedance curve for a conical pipe is shown in figure (*c*). Notice that the peaks and troughs occur together at almost exactly the same frequency; thus it no longer

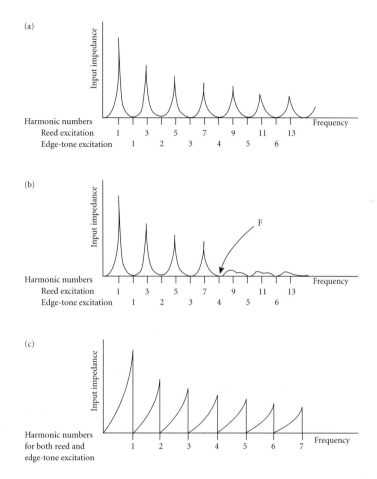

(a)

Input impedance

Harmonic numbers | Frequency
Reed excitation 1 3 5 7 9 11 13
Edge-tone excitation 1 2 3 4 5 6

(b)

Input impedance

F

Harmonic numbers | Frequency
Reed excitation 1 3 5 7 9 11 13
Edge-tone excitation 1 2 3 4 5 6

(c)

Input impedance

Harmonic numbers
for both reed and 1 2 3 4 5 6 7 | Frequency
edge-tone excitation

(*a*) Curve showing the relationship between input impedance and frequency for a plain cylindrical pipe.
(*b*) As (*a*), but with a regular series of side holes.
(*c*) Input impedance plotted against frequency for a conical pipe.

matters whether the pipe is excited by edge tones or reeds, and the full series of harmonics is always produced.

Such diagrams have also been used by Benade to explain the behaviour of trumpets, which seem to produce a full series of harmonics in spite of being largely cylindrical and closed by the player's mouth. Changes are produced in the input impedance curve, first by the addition of the bell and then by the addition of the mouthpiece, and the result is a full sequence of harmonics.

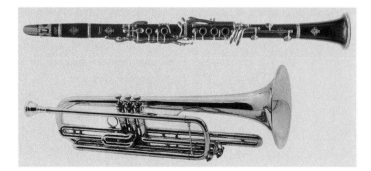

Clarinet and Trumpet.

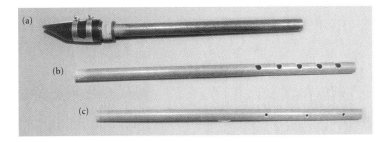

Three plastic tubes used to illustrate the influence of the finger holes on the tone quality of a clarinet: in (a), the tube has no side holes and the tone is muffled; in (b), there are five large holes and the tone is much more clarinet-like; in (c), there are three small holes and the tube is unplayable; however, if one or more of the holes are covered, a note can be sounded.

Turning a cylindrical tube into a clarinet

Just as the trumpet involves departures from a plain cylindrical tube, so does a clarinet. The important departures are the side holes which, even when closed, produce regular 'bumps' in the bore, and these have a profound effect. There are at least three functions that have to be performed by the side holes if a clarinet is to behave like a real musical instrument. The first, and most obvious, is that they determine the vibrating length, and hence the pitch of a given note. Secondly, they radiate the sound (very little of which emerges from the bell, as can be demonstrated easily with a microphone and oscilloscope) and, being arranged in a regular sequence, are frequency-sensitive. Thirdly, there must be a balance between the energy reflected back towards the reed to keep the oscillation going, and the energy radiated away. The position and spacing of the holes has a considerable influence on this. A clarinet maker must be able to adjust at least these three functions independently and, in order to do this, makes use of the positions of the holes, their diameter, the wall thickness at the hole, and the bore diameter throughout the length. The diameter varies along the whole length and is adjusted either by using a reamer to enlarge it slightly at a particular place, or by using a special brush to paint lacquer on the wall to reduce the bore.

The quality of a musical sound

At quite an early stage in the study of musical physics, it was thought that a vibrating body that could vibrate at a number of discrete harmonic frequencies could probably vibrate in several at once, and that the resulting combination of a number of harmonics could be the source of variations in quality. The wave-forms of various instruments were studied, and it seemed clear that their regular waveforms could be subject to Fourier analysis; thus, if the harmonics could be generated in the right proportions, the sound of any instrument could be imitated. The Hammond and Compton electronic organs, both developed in 1932, used the principle of harmonic mixture to determine tone quality. But, as is now well known, the sound of these organs was noticeably 'electronic', and we need to ask why.

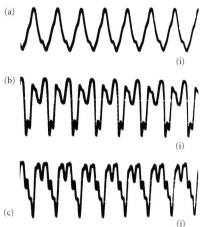

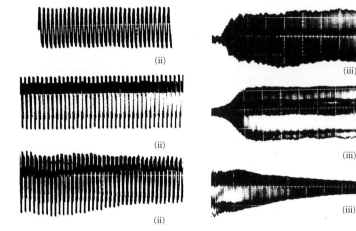

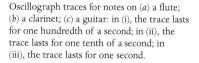

Oscillograph traces for notes on (*a*) a flute; (*b*) a clarinet; (*c*) a guitar: in (i), the trace lasts for one hundredth of a second; in (ii), the trace lasts for one tenth of a second; in (iii), the trace lasts for one second.

The oscillograph traces for three instruments (flute, clarinet and guitar) are obviously different, and over a period of $\frac{1}{100}$ second all three appear to be fairly regular. However if we look at traces lasting $\frac{1}{10}$ second, or 1 second, it immediately becomes obvious that they are far from regular. It turns out that it is these departures from regularity that tell the brain that a 'real' instrument is involved, rather than an electronically synthesized sound. Nowadays, of course, synthesizers have become so sophisticated that departures from regularity can be imitated.

There are many causes of these variations in real instruments, but probably the most significant from the point of view of recognition by the brain is the way in which the note is initiated. Most instruments involve at least two coupled systems: the strings of a violin and the body, the reed of a clarinet and the pipe, the lips of a player and the trumpet itself, and so on. When any coupled system begins to oscillate, one of the systems begins to drive the other in forced vibration. Because of the inertia of the second system, there is a time delay in the commencement of the forced vibration and it may take as much as $\frac{1}{10}$ second before the whole settles down. But this first tenth of a second is crucial: it is called the *starting transient* and every instrument has its own characteristic transient. It is the transient that the brain recognizes, and so permits a listener to identify the various instruments in a combination. Mathematically, the solution of the differential equation for the coupled system is the sum of two parts, the steady state part and the transient part.

Combinations of notes

The phenomenon of beats (rises and falls in amplitude) is well known, and can be demonstrated most easily by sounding the same note on two recorders and then slightly covering the first open hole on one of them to flatten its note slightly. If the two notes have frequencies of 480 Hz and 477 Hz, the beats occur 3 times per second.

Trace for notes of frequencies 480 and 477 Hz, sounded simultaneously.

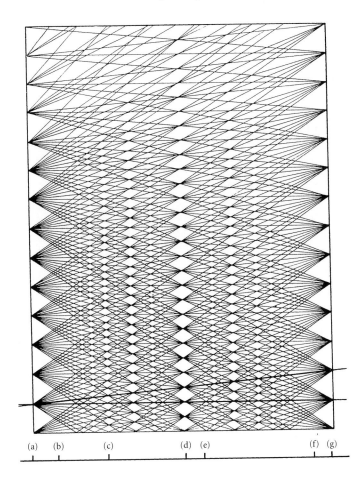

In fact, because of non-linearities in the ear–brain system, a note corresponding in frequency to the difference of the two sounding notes can be heard; this is known as the *difference tone*. The result can be very complicated as there are also sum tones, and there are secondary sum and difference tones between the primary sum and difference tones.

Diagram representing some of the results of adding two pure tones. The horizontal thick line represents a note of fixed frequency; the sloping thick line represents a note whose frequency commences from that of the fixed note and then glides smoothly upwards through one octave. The frequency ratios represented by the lower case letters are: (*a*) 1 : 1; (*b*) 15 : 16; (*c*) 4 : 5; (*d*) 2 : 3; (*e*) 20 : 31; (*f*) 30 : 59; (*g*) 1 : 2.

(a)   (b)   (c)   (d) (e)   (f) (g)

The above diagram shows the result of performing Helmholtz's hypothetical experiment of sounding one note continuously and bringing a second note from being in tune with the steady note to a pitch an octave higher. The thick lines represent the frequencies of the two

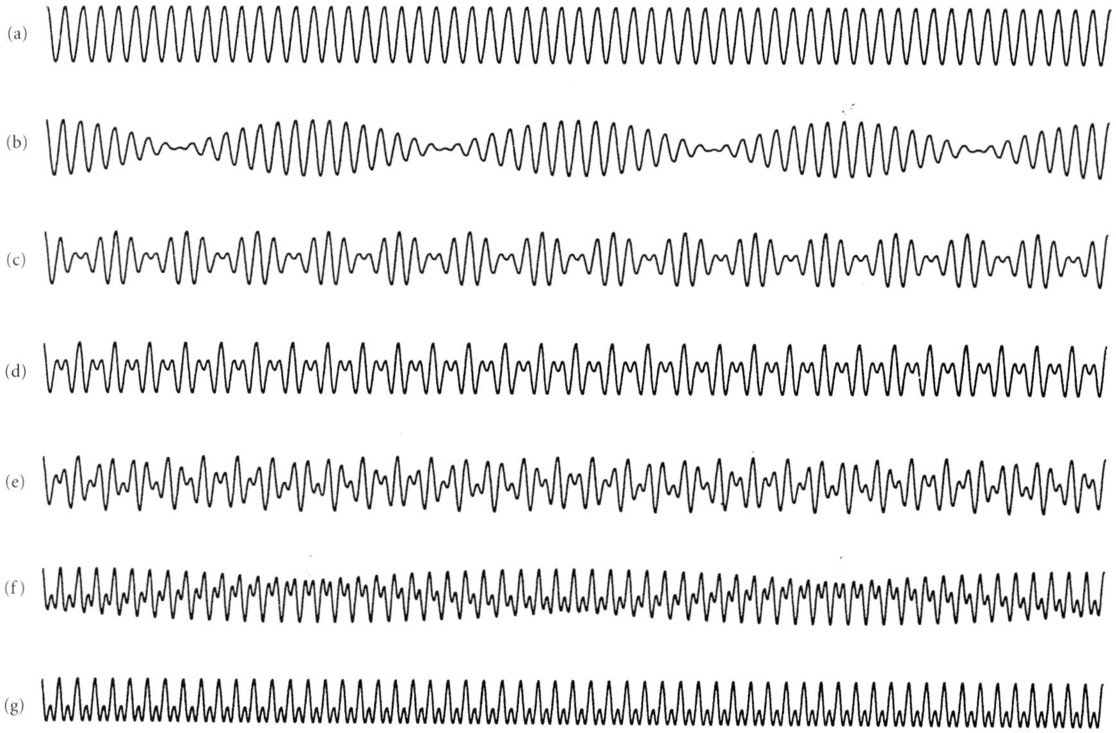

(a)

(b)

(c)

(d)

(e)

(f)

(g)

Wave traces of two simultaneously sounded pure tones; the frequency ratios correspond to those listed in the previous caption.

notes actually being sounded, and the thin lines represent all the various possible sum and difference tones. It can be seen that when the ratio of their frequencies is relatively simple, the number of tones present becomes less. It has been suggested that these tones sound pleasant because fewer notes are involved; also if the wave traces are studied, the simpler ratios give less complicated wave forms. The above diagram shows wave traces for pairs of notes with various frequency ratios.

This looks as though it might begin to account for the phenomena of consonance and dissonance. But there are further complications. The ear–brain system is non-linear only for rather loud sounds, but the sum and difference, and dissonance phenomena, occur even for very low amplitudes.

Also, if three tones are sounded together—say, 400, 480 and 560 Hz— a difference tone at 80 Hz can be heard quite clearly, even at low amplitudes: 80 Hz is the fundamental of the series of which the sounding notes are the 5th, 6th and 7th harmonics. Now, if the frequencies are all raised by the same amount to 420, 500, 580 Hz, although the difference is still 80 Hz, the perceived tone is found to go up by about 10 Hz. The three notes are now the 21st, 25th and 29th harmonics of a fundamental of 20 Hz, although this note cannot be heard. This odd phenomenon, sometimes called the *residue effect*, provides yet one more example of the inadequacy of simple theories to explain musical phenomena. Nor is it an inconsequential complication: the tone of the bassoon, for example, can only be explained using these ideas. If a frequency analysis of bassoon tone is made, there is found to be relatively little energy at the fundamental of any given note. Most of the energy lies in the 5th, 6th and 7th harmonics and the ear–brain system 'manufactures' the fundamental using this residue phenomenon.

The bodies of stringed instruments

Earlier, we mentioned the coupled system incorporating the strings and bodies in the string family. It turns out that the body of an instrument like a violin or a guitar performs an extraordinarily complicated function in transforming the vibrations of the strings into radiated sound. Stradivari, Guarneri, Amati and others obviously solved the problem of making the right kinds of bodies in a purely empirical way and, although physicists can lend assistance to instrument makers in arriving more rapidly at an acceptable solution, the secret of the success of the Cremona school and others is by no means understood. It is clear that much work remains to be done on the science of musical sound.

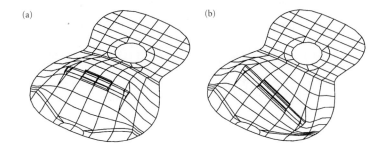

(a)　　　　　　(b)

Computer simulations for two different low-frequency modes of the front plate of a guitar.

Nytt Påfund, til at finna TEMPERATU-REN, i ståmningen för thonerne på Claveret ock dylika Inſtrumenter,

af

DAN. P. STRÅHLE.

Det år vitterligit, at Octaven å Claveret, ock ſlika Inſtrumenter, ei får hafva flera ån 13 thoner: ock om deſſe thoner finge vara aldeles rena, *terſer, quarter* ock *quinter* emellan; ſå fordrar den *Harmoniſka proportionen,* at deras långder, å et *Monochordium,* ſkulle förhålla ſig, ſin emellan, ſåſom nedanſtående upteckning viſar, når *menſuren* för C vore et helt.

C	—	1		fifs	—	$\frac{32}{45}$
Cifs	—	$\frac{15}{16}$		g	—	$\frac{2}{3}$
d	—	$\frac{8}{9}$		gifs	—	$\frac{5}{8}$
difs	—	$\frac{5}{6}$		a	—	$\frac{3}{5}$
e	—	$\frac{4}{5}$		b	—	$\frac{5}{9}$
f	—	$\frac{3}{4}$		h	—	$\frac{8}{15}$
				c	—	$\frac{1}{2}$

T 5

Eller

Faggot's fretful fiasco

Ian Stewart

Musical instruments using fixed intervals, such as pianos and guitars, generally use the equal-tempered scale, so that tunes can be played in different keys. Deciding where to put the frets on a guitar depends, in effect, on finding an approximate construction for the twelfth root of 2. In 1743 a Swedish crafts-man, Daniel Strähle, found a surprisingly simple and good construction, which unfortunately was dismissed by the mathematician Jacob Faggot owing to a mathematical mistake which Faggot made in checking the calculation. The mathematics underlying Strähle's construction is in fact very beautiful and revealing.

If you look at a guitar, mandolin, or lute—any stringed instrument with frets—you'll see that the frets get closer and closer together as the note gets higher. This is a nuisance for the player, because there's less room to fit the fingers in and because the distances you move your finger to get higher notes is not proportional to that for lower ones. But there's a good reason why the frets have to be spaced the way they are: the notes won't sound right otherwise. This is a consequence of the physics of vibrating strings. Today's Western music is based upon a scale of notes, generally referred to by the letters A–G, together with symbols ♯ (sharp) and ♭ (flat). Starting from C, for example, successive notes are

$$\begin{array}{ccccccc}
\text{C}^\sharp & \text{D}^\sharp & & \text{F}^\sharp & \text{G}^\sharp & \text{A}^\sharp & \\
\text{C} & \text{D} & \text{E} & \text{F} & \text{G} & \text{A} & \text{B} \\
& \text{D}^\flat & \text{E}^\flat & \text{G}^\flat & & \text{A}^\flat & \text{B}^\flat
\end{array}$$

and then it all repeats with C, but one octave higher. On a piano the white keys are C D E F G A B, and the black keys are the sharps and flats.

This is a very curious system: some notes seem to have two names, while others, such as B♯, are not represented at all. It is a compromise between conflicting requirements, all of which trace back to the Pythagorean cult of ancient Greece. As we saw in Chapter 1, the Pythagoreans discovered that the intervals between harmonious musical notes can be represented by whole number ratios. They demonstrated this experimentally using a rather clumsy device known as a *canon* (Figure 1), a sort of one-string guitar. The most basic such interval is the *octave*: on a piano it is a gap of

The title page of Daniel Strähle's paper on placing the frets of a stringed instrument.

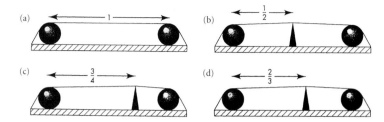

Figure 1. The canon, an experimental device used by the ancient Greeks to study musical ratios: (*a*) full string sounds base note; (*b*) string $\frac{1}{2}$ the length (ratio 2 : 1) sounds note an octave above base note; (*c*) string $\frac{3}{4}$ the length (ratio 4 : 3) sounds note a fourth above base note; (*d*) string $\frac{2}{3}$ the length (ratio 3:2) sounds note a fifth above base note.

eight white notes. On a canon, it is the interval between the note played by a full string (Figure 1a) and that played by one of exactly half the length (Figure 1b). Thus the ratio of the length of string that produces a given note, to the length that produces its octave, is 2 : 1. This is true independently of the pitch of the original note. Other whole number ratios produce harmonious intervals as well: the main ones are the *fourth*, a ratio of 4 : 3 (Figure 1c), and the *fifth*, a ratio of 3 : 2 (Figure 1d). Starting at a base note of C these are

C	D	E	F	G	A	B	C
base			*fourth*	*fifth*			*octave*
1	2	3	4	5	6	7	8

and the numbers underneath show where the names came from. Other intervals are formed by combining these building-blocks.

You can find these ratios on a guitar. Place your left forefinger *very* lightly on the string, and move it slowly along while plucking the string with the right hand. Do *not* depress the string so that it hits any frets. In some positions you'll hear a much louder note. The easiest to find is the octave: place your finger at the middle of the string. The other two places are one third and one quarter along the string.

All guitarists recognise the basic intervals octave, fourth, and fifth. In combination with the fundamental they form the common major chord. A standard 12-bar blues, in the key of C, employs the chord sequence

C/// C/// C/// C/// F/// F/// C/// C/// G/// F/// C/// G///

or a near variant (often with seventh chords instead of major ones in the fourth and final bars). Like this, perhaps:

I got those inhar- monious, Pythagorean blues.　　　*I got those*
　G　　　C/　/　/　C / / /　C/// C⁷///

inhar- monious, Pythagorean blues　　　*Got no*
　F/　/　/　F / / /　C/// C///

har and no mony, an' my guitar's blown a fuse . . .
　G　/　　/　/　　F /　　/　/ C/// G⁷///.

A more traditional song, which comes close, is *Frankie and Johnny*.

It is thought that, in order to create a harmonious scale, the Pythagoreans began at a base note and ascended in fifths. This yields a series of notes played by strings whose lengths have the ratios

$$1 \quad \left(\tfrac{3}{2}\right) \quad \left(\tfrac{3}{2}\right)^2 \quad \left(\tfrac{3}{2}\right)^3 \quad \left(\tfrac{3}{2}\right)^4 \quad \left(\tfrac{3}{2}\right)^5 \quad \text{or} \quad 1 \quad \tfrac{3}{2} \quad \tfrac{9}{4} \quad \tfrac{27}{8} \quad \tfrac{81}{16} \quad \tfrac{243}{32}.$$

Most of these notes lie outside a single octave, that is, the ratios are greater than $\tfrac{2}{1}$. But we can descend from them in octaves (dividing successively by 2) until the ratios lie between $\tfrac{1}{1}$ and $\tfrac{2}{1}$. Then we rearrange the ratios in numerical order, to get

$$1 \quad \tfrac{9}{8} \quad \tfrac{81}{64} \quad \tfrac{3}{2} \quad \tfrac{27}{16} \quad \tfrac{243}{128}.$$

On a piano, these correspond approximately to the notes

C D E G A B.

As the notation suggests, something is missing! The gap between $\tfrac{81}{64}$ and $\tfrac{3}{2}$ sounds 'bigger' than the others. We can plug the gap neatly by adding in the fourth, a ratio of $\tfrac{4}{3}$, which is F on the piano. In fact, we could have incorporated it from the start if we had *descended* from the base note by a fifth, adding the ratio $\tfrac{2}{3}$ to the front of the sequence, and then ascended by an octave to get $2 \cdot \left(\tfrac{2}{3}\right) = \left(\tfrac{4}{3}\right)$.

Figure 2. Scale formed purely from fifths and octaves approximates the white notes on a piano.

The resulting scale corresponds approximately to the white notes on the piano, shown in Figure 2. The last line shows the intervals between successive notes, also expressed as ratios. There are exactly two different ratios: the *tone* $\tfrac{9}{8}$ and the *semitone* $\tfrac{256}{243}$. An interval of two semitones is $\left(\tfrac{256}{243}\right)^2$, or $\tfrac{65536}{59049}$, which is approximately 1.11. A tone is a ratio of $\tfrac{9}{8} = 1.125$. These are not quite the same, but nevertheless two semitones pretty much make a tone. Thus there are gaps in the scale: each tone must be divided up into two intervals, each close to a semitone.

There are various schemes for doing this. The *chromatic scale* starts with the fractions $\left(\tfrac{3}{2}\right)^n$ for $n = -6, -5, \ldots, 5, 6$. It reduces them to the same octave by repeatedly multiplying or dividing by 2, and then places them in order: the result is shown in Figure 3. Each sharp bears a ratio $\tfrac{2187}{2048}$ to the note below it, and from which it takes its name; each flat bears a ratio $\tfrac{2048}{2187}$ to the note above. There's a glitch in the middle: two notes, F$^\sharp$ and G$^\flat$, are trying to occupy the same slot, but differ very slightly from each other.

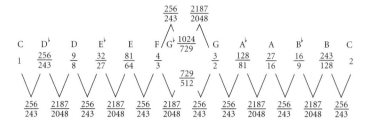

Figure 3. Chromatic scale of twelve notes, incorporating the black notes (sharps and flats); F$^\sharp$ and G$^\flat$, are trying to occupy the same slot.

There are many other schemes, also leading to distinctions between sharps and flats, but they all involve a 12-note scale that is very close to that formed by the white *and* black notes of the piano.

The reason for the glitch in the chromatic scale, and the reason that there are many different schemes for building scales, is that no 'perfect' 12-note scale can be based on the Pythagorean ratios of $\frac{3}{2}$ and $\frac{4}{3}$. A perfect scale is one where the ratios are all the same, so we get

$$1 \quad r \quad r^2 \quad r^3 \quad r^4 \quad \ldots \quad r^{12} = 2$$

for a fixed number r. The Pythagorean ratios involve only the primes 2 and 3: every ratio is of the form $2^a 3^b$ for various integers a and b; for instance, $\frac{243}{128} = 2^{-7} 3^5$. Suppose that $r = 2^a 3^b$ and $r^{12} = 2$. Then $2^{12a} 3^{12b} = 2$, so $2^{12a-1} = 3^{-12b}$. But an integer power of 2 cannot equal an integer power of 3, by uniqueness of prime factorization. Similar arguments show that *no* fixed integer ratio can work.

This mathematical fact puts paid to any musical scale based on Pythagorean principles of the harmony of whole numbers; but it doesn't mean we can't find a suitable number r. The equation $r^{12} = 2$ has a unique positive solution—namely:

$$r = \sqrt[12]{2} = 1.059463094\ldots.$$

The resulting scale is said to be *equally tempered*, or *equitempered*.

If you start playing a Pythagorean scale somewhere in the middle—a change of *key*—then the sequence of intervals changes slightly. Equitempered scales don't have this problem, so they are useful if you want to play the same instrument in different keys. Musical instruments that must play fixed intervals, such as pianos and guitars, generally use the equitempered scale. The Pythagorean semitone interval is $\frac{256}{243} = 1.05349\ldots$, which is close to $\sqrt[12]{2}$, so the name 'semitone' is used for the basic interval of the equitempered scale.

How does this lead to the positions of the frets on a guitar? Think about the first fret along, corresponding to an increase in pitch of one semitone. The length of string that is allowed to vibrate has to be $1/r$ times the length of the complete string. So the distance to the first fret is $1 - (1/r)$ times the length of the complete string. To get the next distance, you just observe that everything has shrunk by a factor of r, so the spaces between successive frets are in the proportions

$$1 \quad 1/r \quad 1/r^2 \quad 1/r^3$$

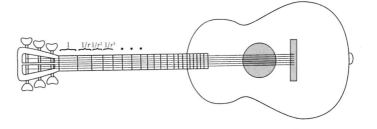

Figure 4. Distances between guitar frets
shrink for the higher notes.

and so on. Now r is bigger than 1, so $1/r$ is less than 1, and that means that the distances between successive frets are *smaller* (see Figure 4).

When the Greeks were faced with numbers such as $\sqrt[12]{2}$ that cannot be written as exact fractions—which they called *irrational numbers*—they usually resorted to geometry. According to tradition, Greek geometry placed considerable emphasis on those lengths that can be constructed using only a ruler and a pair of compasses: for example, squares and square roots can be so constructed (see *Box A*).

The ancient problem of 'duplicating the cube' asks for such a construction for $\sqrt[3]{2}$. This problem is traditionally grouped together with two other problems: three left-overs from Greek geometry, which ask for constructions, using only an unmarked ruler and a pair of compasses, for:

(a) a square whose area is the same as that of a given circle;
(b) an angle one third the size of a given angle;
(c) the side of a cube that is twice the volume of a given cube.

They are known, respectively, as the problems of *squaring the circle*, *trisecting the angle*, and *duplicating the cube*. The transcendence of π, proved by Ferdinand Lindemann in 1882, proves that it is impossible to

---

Box A: Construction of squares and square roots

Constructing squares and square roots with ruler and compasses, given a line of unit length.

Squares: Draw a right triangle AOB with $OA = 1$, $OB = x$. Find the midpoint M of AB and draw MC perpendicular to AB to meet the extension of AO at C. Draw a semicircle with centre C through A, to meet the extension of AO at P. Then OP has length x^2.

Square roots: Draw a line AOB with $OA = 1$, $OB = x$. Find the midpoint M of AB and draw the semicircle centre M through B and A. Draw OP perpendicular to AB to cut the semicircle at P. Then OP has length $\sqrt{x}$.

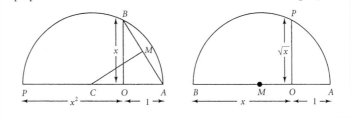

square the circle. The other two problems are also insoluble under the stated restrictions. Before I explain why, it is worth examining the history of these problems in a little more detail, because the usual version of the story tends to produce serious misconceptions.

The first point to make is that the ancient Greeks could solve all three problems. They knew how to find π, they knew about $\sqrt[3]{2}$, and they encountered no difficulty, either theoretical or practical, in splitting an angle into three equal bits—not if they respected the restriction to an unmarked ruler and a pair of compasses, of course, but those restrictions were not widely adhered to by the great mathematicians of ancient Greece. Eutocius, a commentator from the 6th century AD, describes a dozen different methods for duplicating a cube. Several are based on so-called *neusis constructions*, which involve sliding a *marked ruler*—a ruler with a single distinguished point on its edge—along some configuration of lines until some particular condition holds. Others make use of conic sections; and there is a stunning three-dimensional construction that makes use of a cylinder, a cone, and a torus. There are fewer reports of methods for trisecting angles, possibly because there exists an extremely simple and obvious neusis construction (see *Box B*).

A general method for dividing angles by any whole number whatsoever, attributed to Hippias, makes use of a transcendental curve called the *quadratrix*. The quadratrix can also be used to square the circle; and Archimedes, in a surviving fragment that exists only in a much later edition, proves a result which in current terminology states that π lies

---

Box B: Neusis construction for trisecting an angle

To trisect the angle *ABC* draw a circle with centre *O* whose radius is equal to the shaded length marked on the ruler. Draw *OD* parallel to *BC*. Lay the ruler to make *PQ* = *OB*; then triangles *OBP* and *POQ* are isosceles. If angle *POQ* = θ; then angle *PQO* = θ. Then angle *OPB* = 2θ, so angle *OBP* = 2θ. Also angle *PBC* = θ, so angle *ABC* = angle *OBC* = 3θ. Therefore, angle *PBC* trisects the angle *ABC*.

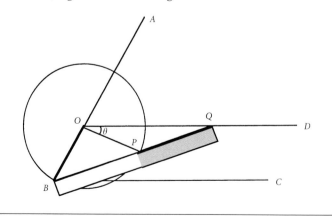

between $3\frac{10}{71}$ and $3\frac{1}{7}$. David Fowler argues that Archimedes may have been trying to evaluate the continued fraction of π, and that the Greeks used continued fractions fairly systematically as a way of forming hypotheses on the rationality or irrationality of particular numbers.

A *continued fraction* is an expression of the form

$$a_0 + \cfrac{1}{a_1 + \cfrac{1}{a_2 + \cfrac{1}{a_3 + \dots}}}$$

which we (mercifully) abbreviate to $[a_0; a_1, a_2, a_3, \dots]$. If Fowler is correct, Archimedes had got as far as $[3; 7, \text{something} \geq 10]$. Had he got as far as $[3; 7, 15, 1, \text{something} \geq 200]$, he might well have begun to wonder whether π might actually be rational, for such unusually large terms normally appear in incomplete developments of rationals.

At any rate, the master mathematicians of classical Greece were perfectly happy using marked rulers and transcendental curves if they needed them. Where, then, did the restriction to constructions with unmarked ruler and compasses come from? It appears to be Eutocius, a mere hack, who explicitly and not very politely criticised the great mathematicians of earlier times for not respecting restrictions that he imposed several centuries later. In short, the now notorious problems 'bequeathed to us by the *ancient* Greeks' were not actually problems that the ancient Greeks, the greats, the *real* mathematicians, ever worried about. It is not unusual for the history of mathematics to be rewritten in this manner.

Be that as it may, later mathematicians took Eutocius's restrictions seriously, and wondered whether constructions might exist that obeyed them. Eventually they proved that there are none, by invoking algebraic methods. Any geometric construction that obeys Eutocius's restrictions can be broken down into elementary steps, and interpreted as a series of solutions to linear or quadratic equations. Therefore any length that can be constructed in the prescribed manner must solve a polynomial equation—indeed, one of a fairly special kind. Duplicating the cube amounts to solving the cubic equation $x^3 = 2$, and this cannot be reduced to a series of quadratics. It follows that there is no ruler-and-compass construction for $\sqrt[12]{2}$ either. For if there were a ruler-and-compass construction for $\sqrt[12]{2}$, then by squaring twice (using ruler and compasses as in *Box A*) we could construct $\sqrt[3]{2}$, which we know is impossible. So there can be no ruler-and-compass construction for $\sqrt[12]{2}$.

The equitempered scale is a compromise, an approximation. The true fourth sounds more harmonious than the equitempered fourth, and singers find it more natural. Since the equitempered scale is a compromise, we may ask whether there is some approximate geometrical construction that tells you where to put the frets on a guitar. Not only is there an approximate construction, but it has a very curious history. The story illustrates the deep elegance of mathematics, but it is also a humbling tale: an outstanding triumph of a practical man nullified by a professional mathematican's carelessness.

In the sixteenth and seventeenth centuries, finding geometrical methods for placing frets upon musical instruments—lute and viol, rather than guitar—was a serious practical question. In 1581 Vincenzo Galilei, the father of the great Galileo Galilei, advocated the approximation

$$\frac{18}{17} = 1.05882\ldots.$$

This led to a perfectly practical method, in common use for several centuries. In 1636 Marin Mersenne, a monk better known for his prime numbers of the form $2^p - 1$, approximated an interval of four semitones by the ratio $2/(3 - \sqrt{2})$. Taking square roots twice, he could then obtain a better approximation to the interval for one semitone:

$$\sqrt{\sqrt{(2/(3 - \sqrt{2}))}} = 1.05973\ldots,$$

which is certainly close enough for practical purposes. The formula involves only square roots, and thus can be constructed geometrically as in *Box A*. However, it is difficult to implement this construction in practice, because errors tend to build up. Something more accurate than Galilei's approximation, but easier to use than Mersenne's, was needed.

In 1743 Daniel Strähle, a craftsman with no mathematical training, published an article in the *Proceedings of the Swedish Academy* presenting a simple and practical construction. Let *QR* be 12 units long, divided into 12 equal intervals of length 1. Find *O* such that *OQ = OR =* 24. Join *O* to the equally spaced points along *QR*. Let *P* lie on *OQ* with *PQ* 7 units long. Draw *RP* and extend it to *M* so that *PM = RP*. If *RM* is the fundamental pitch and *PM* its octave, then the points of intersection of *RP* with the 11 successive rays from *O* are successive semitones within the octave—that is, the positions of the 11 frets between *R* and *M* (see Figure 5).

For practical purposes, Strähle realised that (by similar triangles) a single diagram could be employed with finger-boards of different lengths (see Figure 6).

You might like to try it out, and compare with measurements from an actual instrument. But how accurate is it? The geometer and economist Jacob Faggot performed a trigonometric calculation to find out, and appended it to Strähle's article, concluding that the maximum error is 1.7%. This is about five times more than a musician would consider acceptable.

Faggot was a founder member of the Swedish Academy, served for three years as its secretary, and published eighteen articles in its *Proceedings*. In 1776 he was ranked as number four in the Academy: Carl Linnaeus, the botanist who set up the basic principles for classifying animals and plants into families and genera, was ahead of him in second place. So when Faggot declared that Strähle's method was inaccurate, that was that; for example, F. W. Marpurg's *Treatise on musical temperament* of 1776 lists Faggot's conclusion without describing Strähle's

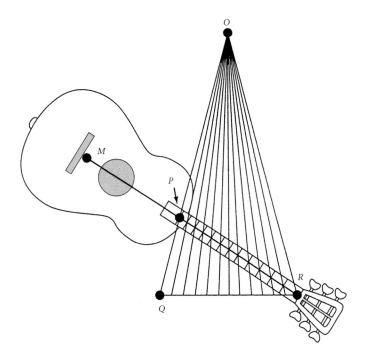

Figure 5.   Strähle's construction.

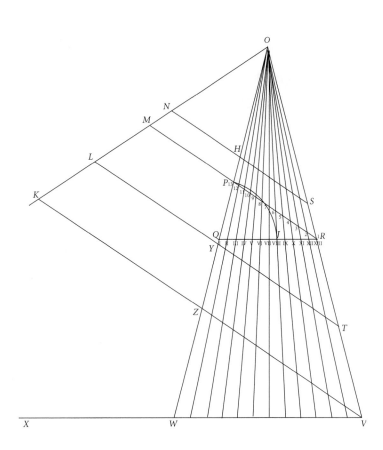

Figure 6.   Strähle's illustration of the practical application of his method. Lay the fingerboard parallel to the line *RPM* and adjust to length, with the midpoint on the line *OW*; now mark the frets.

1743. Octob. Nov. Dec. 287

1. Emedan QR är delt i 12 lika delar, så är hvar deltagen för 1000; då är QR = 12000, QO = 24000, ock QP = QJ = 7000. Men *ángelen* O Q R eller O,Q,VII finnes då

		o	ʼ
blifva - - - -		75.	31
ock som *Angl.* Q,VII, O är - -		90.	—
så blifver *Angl.* Q,O,VII = = -		14.	29
		180.	—

2. Uti *triangelen* O,Q,VII äro nu *ánglarne* ock två sidor bekante, derföre blir sidan OVII = 23240.

Uti *triangelen* PQR äro nu två sidor ock *an-gelen* vid Q bekanta; så blifver

			o	ʼ
angelen - - QPR -			64.	15
ock *angelen* - - PRQ -			40.	14
ty *Angl.* O,Q,VII eller				
OQR eller = = PQR -			75.	31.
			180.	—

4. Emedan nu alla *ánglarna* ock två sidor uti *triangelen* PQR äro bekanta; så blifver sidan PR = 12898.

Uti *triangelen* O, II, VII äro nu två sidor ock den råta *angelen* vid VII bekanta; så blifver

ange-

Figure 7. Faggot's fretful fiasco: the angle *PRQ* is not 40°14′.

method. It was not until 1957 that J. M. Barbour of Michigan State University discovered that Faggot had made a mistake.

Faggot began by finding the base angle *OQR* of the main triangle: it is 75°31′. From this he could find the length *RP* and the angle *PRQ*. Each of the eleven angles formed at the top of the main triangle by the rays from the base could also be calculated without difficulty: it was then simple enough to find the lengths cut off along the line *RPM*.

However (see Figure 7) Faggot had computed the angle *PRQ* as 40°14′, when in fact it is 33°32′. This error, as Barbour puts it, 'was fatal, since [the angle] *PRQ* was used in the solution of each of the other triangles, and exerted its baleful influence impartially upon them all'. The maximum error reduces from 1.7 to 0.15%, which is perfectly acceptable. Thus far the story puts mathematicians, if not mathematics itself, in a bad light: if only Faggot had bothered to *measure* the angle *PRQ*.

But Barbour went further, asking *why* Strähle's method is so accurate. He found a beautiful illustration of the ability of mathematics to lay bare the reasons behind apparent coincidences. There is no suggestion that Strähle himself adopted a similar line of reasoning: as far as anyone knows his method was based upon the intuition of the craftsman, rather than any specific mathematical principles.

The spacing of the *n*th fret along the line *MPR* can be represented on a graph (see Figure 8). We take the *x*-axis of the graph to be the line *QR* in Figure 6, with *Q* at the origin and *R* at 1. We move *MPR* so that it forms the *y*-axis of the graph, with *M* at the origin, *P* at 1, and *R* at 2. The successive frets are placed along the *y*-axis at the points $1, r, r^2, \ldots, r^{11}, r^{12} = 2$. (Note that this differs from the ratios $1/r, 1/r^2, \ldots$ mentioned above, because we are working from the opposite end of the string.)

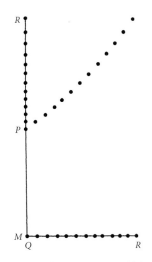

Figure 8.   Graph representing Strähle's construction as a function.

A mathematician would call Strähle's construction a *projection* with *centre O* from a set of equally spaced points along *QR* to the desired points along *MPR*. It can be shown that such a projection always has the algebraic form

$$y = (ax + b) / (cx + d),$$

where *a, b, c, d* are constants: this is called a *fractional linear function*.

For Strähle's method, you can check that the constants are $a = 10$, $b = 24$, $c = -7$, $d = 24$, so the projection takes a given point x on *QR* to the point

$$y = (10x + 24) / (-7x + 24)$$

on *MPR*. I'll call this formula *Strähle's function*. Strähle didn't derive it: it's just an algebraic version of his geometrical construction. However, it is the key to the problem.

If the construction were exact, we would have $y = 2^x$. Then the thirteen equally spaced points $x = n/12$ on *QR*, where $n = 0, 1, 2, \ldots, 12$, would be transformed to the points $2^{n/12} = (2^{1/12})^n = r^n$ on *MPR*, as desired for exact equal temperament. But it's not exact, even though Barbour's calculations show that it's very accurate. Why? The clue is to find the best possible approximation to 2^x, valid in the range $0 \le x \le 1$, and of the form $(ax + b) / (cx + d)$.

One way to do this is to require the two expressions to agree when $x = 0, \frac{1}{2}$, and 1 (see Figure 9). That gives three equations to solve for *a, b, c, d*; namely,

$$b/d = 1; \quad (\tfrac{1}{2}a + b) / (\tfrac{1}{2}c + d) = 2^{1/2}; \quad (a + b) / (c + d) = 2.$$

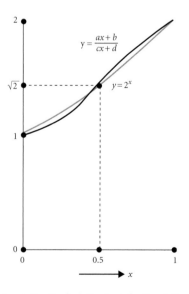

Figure 9. To find the best fractional linear approximation, fit it to the points $x = 0, \frac{1}{2}$ and 1.

At first sight we seem to need one more equation to find four unknowns, but really we only need the ratios $b/a, c/a$, and d/a, so three equations are enough. We may fix the value of d to be anything non-zero, and we decide to set $d = \sqrt{2}$. Then the first equation implies that $b = \sqrt{2}$ as well, while the other two equations simplify to

$$(\tfrac{1}{2}a + \sqrt{2}) / (\tfrac{1}{2}c + \sqrt{2}) = \sqrt{2}; \quad (a + \sqrt{2}) / (c + \sqrt{2}) = 2.$$

Solving these gives $a = 2 - \sqrt{2}$ and $c = 1 - \sqrt{2}$. It follows that the best possible approximation (in our chosen sense) to 2^x by a fractional linear function takes the form

$$y = \frac{(2 - \sqrt{2})\, x + \sqrt{2}}{(1 - \sqrt{2})\, x + \sqrt{2}}.$$

That doesn't look much like Strähle's function, but now comes a final bit of nifty footwork. Barbour estimated the error in terms of the approximation $\frac{58}{41}$ to $\sqrt{2}$, and derived Strähle's formula that way; Isaac Schoenberg did the same in 1982. If you just substitute $\frac{58}{41}$ for $\sqrt{2}$ in the above formula, then you get

$$(24x + 58) / (-17x + 58),$$

which is different from Strähle's function.

Nevertheless, the most natural thing to do is change $\sqrt{2}$ to some approximation—but not $\frac{58}{41}$. Here's how. There is a sequence of rational numbers that approximates $\sqrt{2}$. One way to get it is to start from the equation $p/q = \sqrt{2}$ and square it to get $p^2 = 2q^2$. Because $\sqrt{2}$ is irrational, you can't find integers p and q that satisfy this equation (or, more accurately, because you can't find integers p and q that satisfy this equation, $\sqrt{2}$ must be irrational). But you can come close by looking for integers p and q such that p^2 is close to $2q^2$. The best approximations are those for which the error is smallest—that is, solutions of the equation $p^2 = 2q^2 \pm 1$. For example, $3^2 = 2 . 2^2 + 1$, and $\frac{3}{2} = 1.5$ is moderately close to $\sqrt{2}$. The next case is $7^2 = 2 . 5^2 - 1$, leading to $\frac{7}{5} = 1.4$, which is closer. Next comes $17^2 = 2 . 12^2 + 1$, yielding the approximation $\frac{17}{12} = 1.4166\ldots$, closer still. You can go on forever: to see how, consider the continued fraction for $\sqrt{2}$. Start with the identity

$$\sqrt{2} = 1 + \frac{1}{1 + \sqrt{2}}$$

and then substitute the right-hand side into itself in place of $\sqrt{2}$ to get

$$\sqrt{2} = 1 + \cfrac{1}{1 + 1 + \cfrac{1}{1 + \sqrt{2}}} = 1 + \cfrac{1}{2 + \cfrac{1}{1 + \sqrt{2}}}$$

Repeating the process, we see that $\sqrt{2} = [1; 2, 2, 2, 2, \ldots]$.

If we truncate the continued fraction at some finite position, we get a rational approximation to $\sqrt{2}$. The theory of continued fractions tells us that this must be the best possible rational approximation (with a given size of denominator), and not surprisingly we get a rational $\frac{p}{q}$ with $p^2 = 2q^2 \pm 1$. For example,

$$[1; 2] = \tfrac{3}{2}, \ [1; 2, 2] = \tfrac{7}{5}, \ [1; 2, 2, 2] = \tfrac{17}{12}, \ [1; 2, 2, 2, 2] = \tfrac{41}{29},$$

and so on. We recognise the first three approximations; and for the fourth we find that $41^2 = 2.29^2 - 1$. Indeed, if we write

$$[1; 2, \ldots (n \text{ copies}) \ldots, 2] = p_n/q_n,$$

then

$$p_n/q_n = 1 + \cfrac{1}{1 + [2, \ldots, 2]} = 1 + \cfrac{1}{1 + p_{n-1}/q_{n-1}} = \frac{2q_{n-1} + p_{n-1}}{q_{n-1} + p_{n-1}}.$$

Comparing numerators and denominators, we obtain a pair of recurrence relations

$$p_n = 2q_{n-1} + p_{n-1}; \quad q_n = q_{n-1} + p_{n-1}.$$

For example, from $p_3 = 17$, $q_3 = 12$ we generate

$$p_4 = 2 \cdot 12 + 17 = 41; \quad q_4 = 12 + 17 = 29.$$

Continuing this process we get a table of approximations:

n	1	2	3	4	5	6	7	8	9	10
p_n	3	7	17	41	99	239	577	1393	3363	8119
q_n	2	5	12	29	70	169	408	985	2378	5741

Here, each successive q_n is the sum of the two numbers in the previous column; each p_n is twice the lower number plus the upper number in the previous column. So we have a quick and efficient way to generate rational approximations to $\sqrt{2}$, and incidentally we have proved that the Diophantine equation $p^2 = 2q^2 \pm 1$ has infinitely many integer solutions. Pursuing these ideas leads to a beautiful theory of the so-called *Pell equation* $p^2 = kq^2 \pm 1$. In fact, it was Lord William Brouncker, and not John Pell, who developed the theory: the ideas were erroneously attributed to Pell by Leonhard Euler.

At any rate, we now have lots of approximations to $\sqrt{2}$, among them being $\frac{17}{12}$. Now back to Strähle's equation:

$$y = \frac{(2 - \sqrt{2}) x + \sqrt{2}}{(1 - \sqrt{2}) x + \sqrt{2}}.$$

Divide the numerator and denominator by 2 and rewrite it as the equivalent formula:

$$\frac{x + (1 - x)/\sqrt{2}}{\tfrac{1}{2}x + (1 - x)/\sqrt{2}}.$$

Then replace $\sqrt{2}$ by the approximation $\frac{17}{12}$, so that $1/\sqrt{2}$ becomes $\frac{12}{17}$: this gives

$$\frac{x + \frac{12}{17}(1-x)}{\frac{1}{2}x + \frac{12}{17}(1-x)}.$$

This simplifies to give

$$\frac{10x + 24}{-7x + 24},$$

which is *precisely* Strähle's formula!

So Strähle's construction is very accurate because it effectively combines two good approximations:
the best fractional linear approximation to 2^x is

$$\frac{(2-\sqrt{2})x + \sqrt{2}}{(1-\sqrt{2})x + \sqrt{2}},$$

and Strähle's function is then obtained from this formula by replacing $\sqrt{2}$ by the excellent approximation $\frac{17}{12}$.

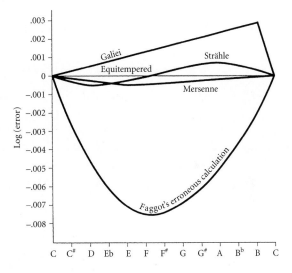

Figure 10. Errors in various constructions: the size of the error is measured by taking the logarithm of the ratio of the approximate value to the true value.

The errors corresponding to the various approximations discussed above are compared in Figure 10: the biggest errors are Faggot's. Thanks to the mathematico-historical detective work of Barbour, we now know not only that Strähle's method is extremely accurate. We also have a very good idea of *why* it's so accurate: it's related to basic ideas in approximation theory and in number theory.

This leaves just one question unanswered—and, barring a miracle or time travel, unanswerable. *How on earth did Strähle think of his construction to begin with?*

CHAPTER 5

Helmholtz: combinational tones and consonance

David Fowler

Not until the 19th century were scientists able to draw together evidence and techniques from acoustics, physiology, physics, technology, psychology, anatomy, and mathematics in order to begin to answer long-standing questions about musical phenomena. Two such questions to which Hermann von Helmholtz (1821–94) made a major contribution concerned combinational tones (where two tones sounding together produce a third and more) and the age-old fundamental Pythagorean insight into consonance.

A question that has teased the minds of musicians and listeners ever since Greek times is: *why* is there an association between musical sounds and simple whole number ratios? What explanation can there be for the Pythagorean insight that consonances seem to relate to small whole numbers and their ratios? Plato in the fourth century BC wrote of the need to 'investigate which numbers are concordant and which are not, *and why each are so'*. Despite the attention of some very distinguished mathematicians and musicians down the centuries—including Kepler, Galileo, Stevin, Bacon, Descartes, Gassendi, Mersenne, Euler, Tartini, d'Alembert, and others—this question did not receive a satisfactory answer for more than two thousand years, when a plausible account was given by the great German physiologist and physicist Hermann von Helmholtz. This chapter is devoted to explaining some of his ideas in this area.

Helmholtz's book *On the sensations of tone as a physiological basis for the theory of music*, first published in 1863, must surely be the single most comprehensive, sustained and profound contribution to musical acoustics. It is still in print, in the brilliant, quirky, opinionated English translation by Alexander Ellis, who added to it many notes and appendices of his own. The second edition of this translation (1885) gives a panoramic view of the subject in the heyday of the mahogany, brass, and glass era; one can scarcely imagine what advances Helmholtz, Ellis, and their contemporaries might have been capable of had they had access to microelectronics, linear microphones, oscilloscopes, and modern optics.

The keyboard of Bosanquet's enharmonic harmonium, constructed around 1876, with 53 divisions to each octave. Its operation and use are described in detail by Helmholtz in his pioneering book *On the sensations of tone.*

Two of the many topics Helmholtz treats will convey the sweep and power of his approach. We begin with a phenomenon first described in the eighteenth century.

Combinational tones

Combinational tones are additional notes that you can sometimes hear when two other notes are played together. The question is, why and under what conditions does this happen? To understand the phenomenon better, Helmholtz carried out experiments with sound generators.

The main sound generators of Helmholtz's time—that is, instruments to produce tones for acoustical research, rather than for musical listening—were tuning forks (for notes of fixed pitch), harmoniums (or, more accurately, reed organs), and the siren (convenient for producing sounds of different pitch). Reed organs were used mainly for demonstrations: Helmholtz's translator A. J. Ellis designed one for the purpose, calling it a 'Harmonical', of which there is a surviving example at the Bate Collection in Oxford—it looks just like what it is, a small harmonium—and Helmholtz's book includes a description of the operation and use of Bosanquet's enharmonic harmonium.

Helmholtz's illustration of a double siren is annotated: 'constructed by the mechanician Sauerwald in Berlin'. (Helmholtz was always scrupulous in giving credit to the instrument workers whose skills made his work possible, and also to King Maximilian of Bavaria and other patrons who gave the money for particularly elaborate pieces of apparatus.) There is a careful description of it in his book; there are two sirens mounted on the same shaft, which can produce two tones of variable pitch but fixed relationship, such as unison, octave, fifth, etc. The cylinder of the top can be rotated independently by a handle; with this, one can investigate beats (by rotating the handle at a constant speed), or demonstrate that the ear cannot detect phase shift (by fixing the handle at different settings). We are no longer used to thinking of sirens as scientific instruments, but some still exist as public fire alarms where they are usually double. Next time you hear one, listen carefully: along with the two or more loud tones of the siren itself, you will clearly hear a lower tone, and you may also hear a higher tone. These are *combinational tones*. Let Helmholtz explain them in his own words (note the opening phrase: 'these tones *are heard*'):

These tones are heard whenever two musical tones of different pitches are sounded together, loudly and continuously ... Combinational tones [also known as *terzi suoni*, grave harmonics, resultant tones, subjective tones, intermodulation tones, aural harmonics, and heterodyne components!] are of two kinds. The first class, discovered by Sorge [a German organist, in 1745] and Tartini [the Italian violinist, in 1754], I have termed *differential tones*, because their pitch number [frequency] is the *difference* of the pitch numbers of the generating tones. The second class of *summational tones*, having their pitch number

Helmholtz's illustration of a double siren.

equal to the *sum* of the pitch numbers of the generating tones, were discovered by myself.

The technique for detecting and analysing tones are twofold: either to detect the vibration which produces them, or to pick out the frequency under investigation using tuned resonators. In the first method, Helmholtz might attach a bristle to a tuning fork and draw it across a smoky glass plate; we might photograph the motion of particles, or use optical interferometry, or display the output of a microphone on an oscilloscope. Preyer's tuning forks are an example of tuned resonators; Helmholtz devised a range of glass resonators through which the experimenter would listen. The first method tells us little beyond gross information—it is notorious that the wave-form gives little information about the sound of things—and the ultimate resonator that we must often fall back on to investigate the nature of sound is our own ear. One fascination of musical acoustics is this attempt to extract some objectivity about this essentially subjective phenomenon.

Helmholtz's first interest was physics but, because state aid existed for medical students and his family was of modest means, he trained as a doctor and then started medical practice in the army. However his interest was always more in research and, while a student and in practice, he worked with Johannes Müller and his students and absorbed their philosophy of founding physiology on physical and chemical processes, rejecting ideas of non-physical 'vital forces'. Significantly, Helmholtz's first major paper, in 1847, which arose from his study of the action of muscles, introduced the idea of potential energy and was one of the several simultaneous announcements of the principle of conservation of energy. He was the first to measure the speed of nerve impulses and his result, that it was around 30 metres per second in frogs' nerves, caused great surprise by being so slow. He invented the ophthalmoscope in 1850, and laid down our understanding of the eye in his three-volume *Handbuch der physiologischen Optik* (1856–67). His book *On the sensations of tone* does the same for the ear.

He was among the first to give detailed physiological descriptions of some aspects of the fine structure of the ear, which he summarizes over thirteen pages of his book in a passage that begins: 'The construction of the ear may be briefly described as follows . . .'. The mechanism is exquisitely delicate: the eardrum or tympanum is linked by three small bones (collectively called the ossicles—individually, the hammer, attached to the drumskin, the anvil, and the stirrup) to the inner ear, which is filled with fluid and contains both the balancing mechanism of the semicircular canals and the auditory mechanism of the snail-like cochlea, containing the neural mechanism for detecting and analysing the sound and transmitting the resulting nerve impulses on to the brain.

Helmholtz gives a detailed analysis of all aspects of the ear including, in the spirit of Müller's approach, its mechanical characteristics. For example:

The mechanical problem which the apparatus within the drum of the ear had to solve, was to transform a motion of great amplitude and little force, such as impinges on the drumskin, into a motion of small amplitude and great force, such as had to be communicated to the fluid in the labyrinth. A problem of this sort can be solved by various kinds of mechanical apparatus, such as levers, trains of pulleys, cranes, and the like. The mode in which it is solved by the apparatus in the drum of the ear, is quite unusual, and very peculiar.

One of the peculiarities that Helmholtz described is its asymmetry: the drumskin (or tympanic membrane) is curved inwards, and hammer and anvil are not fixed together but have interlocking teeth that allow a ratchet-like behaviour.

Back, now, to combinational tones. Helmholtz gave a general description of his explanation in the text, but reserved the details to a short appendix. The first-order theory of vibration and hearing is linear, as exemplified by *Ohm's law of perception*:

The human ear perceives pendular vibrations [simple harmonic motions] alone as simple tones, and resolves all other periodic motions of the air into a series of pendular vibrations, hearing the series of simple tones which correspond with these simple vibrations.

This corresponds to the approximation of assuming that all vibrations are infinitesimally small and periodic motion can be resolved into its Fourier expansion; but while, in reality, the vibrations may be very small, they are in no sense infinitesimal. Although Helmholtz does not actually use these words, for they were not then fashionable, his explanation is that combinational tones are the *ear's* non-linear response to these vibrations of finite amplitude. He concludes the short mathematical appendix that gives the detailed argument (in the form of the solution of a differential equation) with these words:

If, then, we assume that in the vibrations of the tympanic membrane and its appendages, the square of the displacements has an effect on the vibrations, the preceding mechanical developments give a complete explanation of the origin of combinational tones. Thus the present new theory explains the origin of the tones $(n + m)$ as well as of the tones $(n - m)$, and shows us, why when the intensities a and b of the generating tones increases, the intensity of the combinational tones, which is proportional to ab, increases in a more rapid ratio ... Now, among the vibrating parts of the human ear, the drumskin is especially distinguished by its want of symmetry, because it is forcibly bent inwards to a considerable extent by the handle of the hammer, and I venture therefore to conjecture that this peculiar form of the tympanic membrane conditions the generation of combinational tones.

Since Helmholtz's time, much delicate work has been done on the fine structure of the ear, but I do not know how the detail of this final hypothesis has stood the test of time. The following quotation from J. F. Bell, a leading experimenter in the field of non-linear elastic phenomena, indicates that the topic may still be far from resolved:

History abounds with unwarranted rejection of valid experiments. Only in hindsight do we learn that a good nonlinear ear is required to hear Hermann Helmholtz' summation tones in musical acoustics, an acoustical property universal among musicians but obviously not a common characteristic of the ears of many, but fortunately not all, physicists since the 1850s. [Bell's footnote: Aural harmonics are a subjective measure of the phenomenon of summation and difference tones. As a small sample of the difference of opinion among physicists, I quote titles from the 'Letters to the Editor' section of a single issue in June 1957 of *The Journal of the Acoustical Society of America*: 'Aural Harmonics are Fictitious'; 'On the Inadequacy of the Method of Beats as a Measure of Aural Harmonics'; 'In Support of an 'Inadequate' Method for Detecting 'Fictitious' Aural Harmonics'.] Only in hindsight, too, do we learn that patience and knowledge, ignored by his numerous contemporary discreditors, were required to reproduce Léon Foucault's pendulum experiment in the mid-nineteenth century.

A photograph of Hermann von Helmholtz taken for Lord Kelvin by Mr Henderson, a student, on 7 July 1894, a few days before Helmholtz's final illness.

At this point, Bell has another footnote:

Foucault presented the results of his experiment to the French Academy on February 3, 1851 and demonstrated the experiment to the general public in the Pantheon in May, 1851. [His pendulum has now been reinstalled there.] Foucault's experiment which demonstrated the rotation of the earth aroused his contemporaries to publish over 60 papers during that same year. There were debates among the theorists who adopted opposing analytical approaches. There was discord among the experimentists, some of whom, not appreciating the demanding requirements of Foucault's experiment on the pendulum, obtained conflicting results. Thus was generated the heated controversy that dominated the remaining 17 years of Foucault's life of only 47 years.

Two comments conclude this section. First, in this model, Helmholtz inferred that the ear generates its own harmonics. So harmonics are all around us, in the physics of most musical instruments, in the mathematics of the analysis of periodic motion, and in the acoustics of our perception of tone. This underlines the second-order violation of Ohm's law of perception: the ear perceives more than the frequencies that objectively are presented to it. Second, there is a simple and obvious explanation of the difference tone that was advanced by Lagrange, Thomas Young, and many others, and was still a live issue at the time of Ellis' translation, that as the beating of two almost coincident pure tones increases in frequency as the tones move apart, so the beating sensation would move into the perception of the difference tone (hence the name 'beat tone' often given to the difference tone). Experiments have been done, with ambiguous results, to see if a beat-like phenomenon can give rise to the perception of a tone, but the most convincing single argument against this proposal—at least to those who can hear it!—is that the summation and other combinational tones cannot be so easily explained. The matter of beating also enters our second topic.

The problem of consonance

One of the oldest problems of science is to explain the Pythagorean association of consonance with small integer ratios. As we saw in Chapter 1, when we sound a unison $1:1$, an octave $2:1$, a twelfth $3:1$, a double octave $4:1$, a fifth $3:2$, or a fourth $4:3$, we get a blended harmonious sound, different in quality from an interval taken at random. Such a repertoire of harmonious sounds has been at the basis of the Western and some other musical traditions, as far as we can trace back in time. So just *what* is consonance, and just *why* are these particular intervals perceived as consonant?

Plato included music alongside mathematics and astronomy in the curriculum for the future rulers of his state that he set out in his

Republic, Book VII, explaining the parallel as follows:

It appears that just as the eyes are fixed on astronomy, so the ears are fixed on harmonic motion, and these two sciences are one another's sisters, as the Pythagoreans say and we agree...[But we must] rise to problems, to investigate which numbers are concordant and which are not, and why each are so.

There is also a Greek treatise (attributed to Euclid) on the division of the scale, the *Sectio canonis*, whose obscure introductory essay concludes:

Among [pairs of] notes we also recognize some as concordant, others as discordant, the concordant making a single blend out of the two, which the discordant do not. In view of this, it is to be expected that the concordant notes, since they make a single blend of sound out of the two, are among those numbers which are spoken of under a single name in relation to one another, being either multiple [of the form $n:1$] or epimoric [of the form $(n+1):n$].

This abstract principle—if concordant, then multiple or epimoric—seems to have mediated the Pythagorean approach to music, for they do not seem to regard the eleventh, the octave plus fourth with ratio $8:3$, as concordant, to the derision of those Greek music theorists who had a closer eye on practice. Here, for example, is what Aristoxenes writes, in the *Elementa harmonica II*:

We must explain first that the addition of any concordant interval to the octave makes the magnitude resulting from them concordant.

The problem of explaining consonance was a live issue until comparatively recent times: Kepler, Galileo, Stevin, Bacon, Descartes, Gassendi, Mersenne, Rameau, Euler, Tartini, d'Alembert, and others all gave explanations of the phenomenon (although not all of them published their thoughts), and many of them regarded themselves as providing the first really satisfactory explanation. For example, Kepler wrote in his *Harmonices mundi* of 1619:

After two thousand years [during which the causes of the intervals] had been sought for, I am the first, if I am not mistaken, to present them with the greatest precision.

and Galileo wrote, in his *Two new sciences* of 1638:

I stood a long time in Doubt concerning the Forms of Consonance, not thinking the Reasons commonly brought by the learned Authors, who have hitherto wrote of Musick, sufficiently demonstrative...We may perhaps be able to assign a just reason whence if it comes to pass, that of Sounds differing in Tone, some Pairs are heard with great Delight, others with less; and that others are very offensive to the ear.

By general consent, Helmholtz's explanation, more than two centuries later, is much ahead of the rest, bestriding as it does all aspects of the problem: instrumental, acoustical, physiological, psychological, and mathematical.

What features should a satisfactory explanation of consonance possess?

- It should centre on some characteristic that we can recognize, something that makes precise what Helmholtz, echoing the *Sectio*

canonis, described in the words 'consonance is a continuous, dissonance is an intermittent sensation of tone'.

- It should be broad enough to admit some change over time, to
  allow for intervals such as the third, the sixth, and the seventh, that
  gradually move into the corpus of permitted consonant intervals;
  or to explain why the fourth $4:3$, a classically consonant interval,
  may not seem as consonant as the third $5:4$, or even the sixth $5:3$.

- It must be broad enough to admit that most of the so-called
  consonant intervals in our music are mistuned: our tempered fifth,
  used in almost everything except unaccompanied choral singing, is
  not $\frac{3}{2}:1$ but $2^{7/12}:1$ (that is $1.4983\ldots:1$), while the major third is
  worse: if the instruments and players are indeed playing in properly
  tuned equal temperament, the third is not $\frac{5}{4}:1$, but $2^{1/3}:1$
  ($=1.2599\ldots:1$). We do not hear nice ratios but musical notes!

- On the other hand, it should explain how some intervals, such as
  the unison and octave, can stand no tempering, while other intervals are much more tolerant.

- Ultimately, it should be able to explain the ambiguity of intervals: that
  a grossly mistuned major sixth may in some circumstances be recognized as a widened major sixth, and in others as an arrowed minor
  seventh, but never as both simultaneously; or an interval of six semitones (a tritone) may be parsed in one context as a diminished fifth,
  and in another as an augmented fourth.

- It must also be broad enough to explain how some intervals such
  as the major third sound more consonant in the treble than in the
  bass. Or, in a celebrated prediction of Helmholtz, that a major third
  D–F♯ played by a clarinet and oboe sounds much better when the
  clarinet takes the lower note than when the oboe does, while a
  fourth or minor third will sound better when the oboe takes the
  lower note.

Having put together such a formidable list of desiderata, let us
embark on Helmholtz's explanation. As has just been suggested, the
instruments involved are significant so let us fix, for example, on two
violins. (This choice is especially favourable, as we can play notes of
every pitch with a violin, so chords of any interval with two of them.)
We know that the violin's quality of tone is produced by its own particular mix of fundamental and overtones, so we do a harmonic analysis
to determine what precisely this mix is.

The listener is crucial, so we do an experiment there also. This time
we use pure tones—and, in this precis of the description (but not in
Helmholtz's book!), we ignore the complications of the combinational
tones and other non-linear responses of the ear. We, the subject, hear
two tones that start in unison; as one of them increases in pitch, we
hear beating—slow at first, then getting quicker, and more unpleasant,

until it reaches its peak of unpleasantness at around 30 beats per second. After this stage the quality of unpleasantness—*roughness* is the word used by Helmholtz—decreases away to zero. So we have some sort of 'unpleasantness curve' such as this:

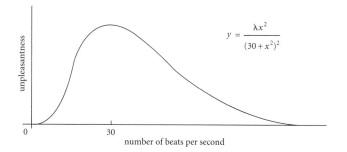

Helmholtz took the simplest such kind of expression, which he admitted was an arbitrary choice, but

it at least serves to shew that the theoretical view we have proposed is really capable of explaining such a distribution of dissonances and consonances as actually occurs in nature.

Take, then, two violins playing any chord, and, using the harmonic analysis of the violin's tone and this roughness curve, add up the contributions of each constituent of the two notes: this gives a total measure of the roughness of the interval. Better, says Helmholtz, 'knowing that diagrams teach more at a glance than the most complicated descriptions', to draw all this out as a graph. His figure appears opposite, split into two halves, with a further scale added. (It is not clear what technology he used to produce his graph, but it is difficult to read and interpret in his book—in particular, it was printed in white on black; our version has been greatly cleaned-up and enhanced.) One violin plays middle C, while the other violin plays any note in the two octaves above.

Look at the left-hand end of the top graph, around the point middle C. The lowest curve gives the roughness of the two fundamentals, labelled 1 : 1, to which is added the roughness of the harmonics, labelled successively 2 : 2, 3 : 3, 4 : 4, and 5 : 5, their magnitudes being determined by the harmonics analysis of the violin's tone. As the interval widens, contributions from the ninth harmonics of the lower note and the eighth of the higher note enter, then of the eighth and seventh, while the contributions of the harmonics of the unison fade away into insignificance, and so on. The sum of all of these contributions is the top-most curve, the total roughness curve. The minima of this total roughness curve then give the points of relative local consonance: we find steep valleys at the most perfect points of consonance, especially the unison, the octave (where the two graphs join), and the twelfth (on the lower half), and less well-defined minima at the imperfect consonances. The steeper the valley, the more critical is the tuning; the

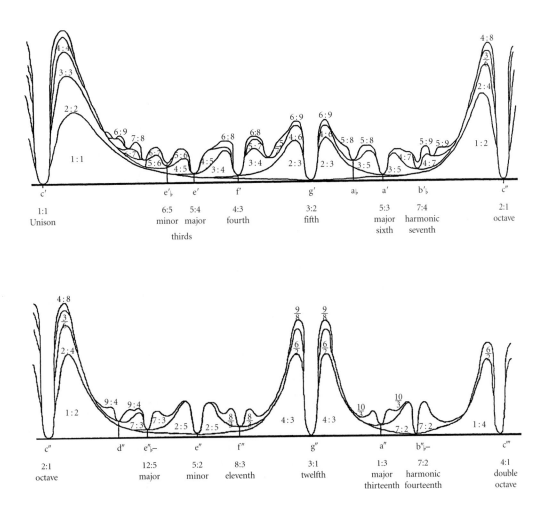

lower the trough, the greater the degree of consonance. His diagram is not definitive: it clearly required an immense amount of labour to calculate this example by hand, and it can be refined. (For an obvious example, the unison involves only contributions up to the fifth harmonic, although data is elsewhere included up to the tenth, so the valley at the unison should be steeper; translated into musical terms, this means that its tuning is even more critical.)

Such was Helmholtz's explanation. With rather more justification than some of his predecessors in the long tradition, he says:

I do not hesitate to assert that the preceding investigations, founded on a more exact analysis of the sensations of tone, and upon purely scientific, as distinct from esthetic principles, exhibit the true and sufficient cause of consonance and dissonance in music.

Concluding remarks

Two things cannot but strike the mathematically alert twenty-first-century reader in this backward look to the nineteenth century. Two fashionable bits of mathematics in recent years have been non-linear analysis (with the possibility therein of chaotic behaviour) and catastrophe theory. Helmholtz's passionate belief was that the non-linear contribution, the second and higher order effects, are essential to an understanding of musical acoustics, and his minimum-seeking approach to consonance is the exploitation of a pure catastrophe-theoretic technique. And while Helmholtz's book is a delight for those who like their science broad and deep, embracing physiology, physics, and mathematics, we have not begun to explore his appreciation of the stuff of sound, music itself. Finally, the reader will find there gems of other diverse sorts; for example, the elaborate connection between the non-conformist church, tonic sol-fa, music publishing, and the nineteenth-century attempt to construct keyboard instruments that could play in just intonation, or at least better approximations to it than that provided by the crude tempered diatonic scale. Read this for yourself!

PART III

Mathematical structure in music

MENUET AL ROVESCIO

CHAPTER 6

The geometry of music

Wilfrid Hodges

The dimensions of time and pitch make music into a two-dimensional space. Geometers study a space by describing its possible transformations, and they study a pattern in space by asking what transformations leave the pattern unchanged—that is, what symmetries the pattern has. We apply these ideas to musical space. For example, when does it make musical sense to squeeze a tune, or to turn it upside down? Since musicians cannot use very high or low pitches, a piece of music is like a frieze; we can find musical example of all the possible symmetries of a frieze pattern.

In memory of Graham Weetman (1963–92), mathematician and musician.

The rise and fall of musical space

While Edward Elgar was writing his *Enigma Variations*, he went for a walk along the banks of the River Wye with his friend G. R. Sinclair. Sinclair brought his bulldog Dan, who fell in the river and barked as he climbed out again. Sinclair turned to Elgar and said 'Set that to music'. So Elgar did, in the variation named *G.R.S.* after Sinclair. Elgar's manuscript marks 'Dan' at the point where Dan barks, though the printed editions rather prudishly leave it out.

Here is the score of the crucial moment:

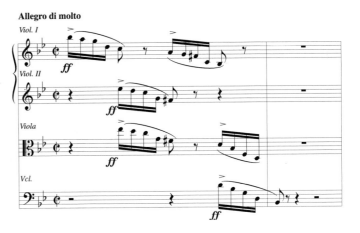

An example of a musical palindrome: the minuet from Joseph Haydn's *Piano sonata No. 41, Hob. xvi/26.*

Edward Elgar, *Enigma variations XI 'G.R.S.'*

You can clearly see the lines slanting down from top-left towards bottom-right. Since the music runs from left to right, these lines represent the music *falling*, and indeed you can hear it falling if you listen to a performance. Elgar makes the music fall because the dog fell. But these are two quite different kinds of falling. In the musical score, high-pitched notes (notes of short wavelength) appear near the top, and low-pitched notes (notes of long wavelength) are near the bottom. So a fall in the score indicates that the orchestra moves from short wavelengths to long wavelengths. The dog, on the other hand, falls by moving rapidly two or three metres closer to the centre of the earth.

Johann Jakob Froberger, *Suite XII in C major,*
Lamento sopra la dolorosa perdita della Real Mstà
di Ferdinando IV, Ré de Romani.

Here is an example of the opposite phenomenon: music that rises to describe something going up. It comes at the end of Froberger's musical depiction of the death of Emperor Ferdinand IV. The picture shows Froberger's manuscript, and you can see Froberger's picture of the clouds of heaven welcoming the soul of the emperor as it climbs up a scale of three octaves.

The two examples above illustrate the difference between up or down in space and up or down in musical pitch. In fact there are not two but three different kinds of space to be correlated. First, there is *physical space*. It has four dimensions—three of space and one of time. Second, there is *the score*. To a first approximation, the score is a plane surface with a horizontal dimension and a vertical dimension. By convention the horizontal dimension represents time, from the past on the left to the future on the right. Also by convention the vertical dimension represents pitch; notes of shorter wavelength are written higher up. Third, there is *musical space*. This space has any number of dimensions, depending on how we choose to analyse it. The two most obvious dimensions are time and pitch, and these are the two that we represent as dimensions in the score. Probably the best candidate for a third dimension is loudness. But the human ear is very bad at comparing the loudness of different sounds, and even worse at remembering degrees of loudness. Most music doesn't distinguish more than five or six degrees of loudness.

Taken literally, time in music just is physical time; some musical events happen before others, and 'before' means 'earlier than' in the usual physical sense. It's pure convention that time is represented in the score by movement from left to right. The same convention makes physicists put time on the *x*-axis, moving forwards from left to right.

Pitch is a more complicated matter. To us today it seems obvious that high-pitched notes are 'high' and low-pitched ones are 'low'. That's how we are able to understand the music of Elgar and Froberger quoted above. So it comes as a shock to learn that in classical Greece high-pitched notes weren't heard as 'high'. In fact the highest-pitched note of the classical Greek octave was called *nete*, the 'nether' or lowest note. It got this name from the fact that the stringed instrument called a kithara was held with the highest-pitched string nearest the ground—just how one holds a guitar today. The classical Greek expression for high-pitched notes was *oxys* 'sharp', whence our musical sharps today; the Greek for low-pitched was *barys* 'heavy'. Classical Greek musicians represented pitches by letters, not by position on a musical page.

Very likely the correlation between short wavelength and height on the musical page was set up before anybody connected either of them with physical height. The correlation was made in western Europe, probably during the period 850–1150 AD. Two manuscripts of the late ninth century use a system of labelled boxes for the pitches, and they both put the boxes for higher pitches nearer the top of the page. This clumsy system never caught on. But during the next few centuries a notation developed for showing where the music rises and falls, and the notation formed a strong tendency to show rises in pitch by shapes like /, and falls in pitch by shapes like \. This led directly to the modern staff notation, which started to emerge in the twelfth century.

Nobody knows just when high pitch came to be correlated with physical height, but the correlation seems to come from western Europe again, and it became very strong during the fifteenth century. This is the century in which the names *altus* or *superius* 'high' and *bassus* 'deep' came to be used for high-pitched and low-pitched voices—whence our altos and basses. (Altos count as high because women weren't allowed to sing in churches.) During this century there were still a few notation systems that put lower-pitched sounds higher on the page—for example, some Italian lute tablatures—but they disappeared at the beginning of the sixteenth century, presumably because they had come to feel too unnatural.

Josquin Desprez put the seal on this development by sending Jesus down from Mount Olympus in a descending scale of twelve notes. This is still one of the longest descending motifs in all vocal music (see overleaf).

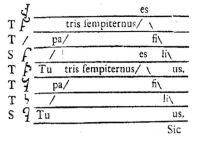

Late ninth-century notation, copies in print by Gerbert, *Scriptores ecclesiastici de musica*, 1784.

Josquin Desprez, *Huc me sydereo* (late 1490s).

Up, down, between and distance

Josquin's idea opened the floodgates to a torrent of musical representations of ups and downs, mostly in music from western Europe. We find them in Byrd, Purcell, Handel, Haydn, Wagner and Elgar.

An interesting example is a moving passage from the chorus *The cold deepens* in Michael Tippett's oratorio *A child of our time*. The first staff is the soprano part and the second staff shows some of the notes played by the orchestra.

Michael Tippett, *A child of our time, No. 26 Chorus*.

While the soprano sings 'The world descends' and duly sinks downwards into the frozen ocean, the orchestral part surprises us by moving upwards. Tippett knows exactly what he is doing. Physical space has more structure in it than just up and down. It also has distance, and as distance changes in time we have moving apart and coming together. We can carry these notions over to musical space. In fact, Tippett's piece conveys a strong feeling that as the sopranos descend and the bass instruments of the orchestra rise to meet them, something is getting trapped between the two.

Franz Schubert, song: *Death and the maiden*.

Tippett was by no means the first composer to play this metaphor. Schubert has a very similar device in his song setting of Claudius' poem *Death and the maiden*. In the first half of the song the maiden begs Death to leave her alone ('Rühre mich nicht an'). She sings energetically, but the bass line in the piano betrays that her strength is sinking. Then suddenly the bass line turns upwards; not because her strength comes back, but because Death is trapping her between the right hand and the left hand of the pianist. From that point onwards, only Death sings.

Charles Ives set himself an impossible problem. He wanted to use pitch distance to represent the fact that God is infinitely close to man. But what is an infinitesimally close pitch distance? In the end Ives gave up and left it to the singer to decide. Maybe what Ives wanted was a smallest perceptible pitch difference. There is no standard notation for this.

Charles Ives, *Duty.*

How does one measure distances of pitch? For classical western music there are two main answers. The first is that pieces of music tend to be in a key, and we count one unit of distance for each step up the scale of that key. This measure of distance is called the *diatonic metric*, and it depends on the key. ('Metric' is the mathematicians' name for a scale of measurement.) The second answer is that for many instruments with set pitches (such as pianos and organs) the smallest distance between two playable notes is a semitone; so we count one unit of distance for each semitone. This is the *chromatic metric*. Thus from B to F is 4 diatonic units in the scale of C major, 3 diatonic units in the scale of F♯ major and 6 chromatic units:

Metrics: diatonic in C major; diatonic in F♯ major; chromatic.

Things become immensely more complicated as soon as one moves even a short distance from western classical music.

There is a natural dual to Ives' question: How can we use musical space to represent that two things are *infinitely far apart*? For some reason, composers have generally wanted to do this more with time than with pitch; the problem is to represent eternity within the confines of a piece that lasts, say, half an hour. There are several ways to do it. A simple way is to make a note last not for ever but for a relatively long time. Thus Wagner in *Parsifal*:

('You and I would be damned for eternity, for the sake of one hour.')

Richard Wagner, *Parsifal*, second act.

Auf E - wig - keit wär'st du ver-dammt mit mir, für ei - ne Stun-de

Here eternity lasts seven crotchets, compared with the two crotchets of an hour, giving the rather unimpressive ratio of 3.5 hours to 1 eternity.

Haydn in *The heavens are telling* from his *Creation* puts two pauses on 'ever', daring Gabriel (the soprano voice) to make them last as long as she can. (In the German version the pauses sit pointlessly on the word 'keiner'; but the libretto for *Creation* was written first in English.)

Joseph Haydn, *Creation*, chorus: *The heavens are telling*.

As we shall see below, other composers have used a subtler and more geometric way of pointing to eternity.

What is space?

Until the middle of the nineteenth century there was very little for mathematicians to say about musical space. This was because mathematicians had a shallow view of space itself; they thought of it as built up from points, lines, planes, and so forth. There is not much to be said about musical points and lines.

But the second half of the 19th century, particularly the work of Felix Klein, brought a new view of what space is. Instead of asking what space is made of, we ask what are the significant *transformations* of space. Roughly speaking, a transformation of space is a rearrangement of space and the things in it, that can be written by a simple mathematical formula.

Our musical space has just the two dimensions of pitch and time, so it forms a two-dimensional space—in fact, a plane. The figures below illustrate four kinds of transformation of a plane. The light horn is a set of points of the plane; the transformation moves the plane so that those points finish up forming the dark horn. The transformed version of a set of points in the plane is called the *image* of the set; so the dark horn is the image of the light one.

Felix Klein (1849–1925)

- A *translation* is a transformation T that moves all points of the plane in the same direction and through the same distance. (If the distance is 0, then *T* is the *identity transformation* that leaves everything exactly as it was.)

Translation

- A *rotation* is what the name suggests: it rotates the whole plane through some angle strictly between 0° and 360° around some fixed point. (In the illustration the fixed point is in the middle of the circle made by the horn's tubes, and the rotation is clockwise, as shown by the arrow.)

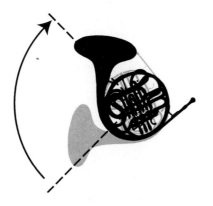

Rotation

- A *reflection* is what you get if you put a two-sided mirror at right angles to the plane and take each point to its reflection in the mirror. The dashed line shows the mirror.

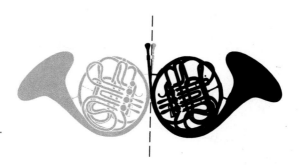

Reflection

- A *glide reflection* is the hardest to describe. This transformation consists of a translation along a line (the dashed line of the illustration), followed by a reflection in the same line.

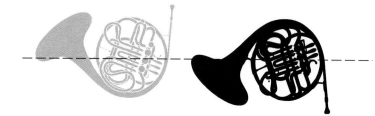

Glide reflection

These four kinds of transformation have an important property: they never change the distance between any two points on the plane. (This is sometimes expressed by saying that they move the plane 'rigidly'.) Transformations with this property are called *isometries* because they don't alter the scale of distances (the metric). There is a theorem of geometry which tells us that every isometry of the plane has one of the four types above.

We can apply these ideas to musical space as follows. At a first approximation, a musical motif *M* consists of a set of notes performed at certain pitches over certain time intervals; so it is a subset of musical space. A *symmetry* of *M* is an isometry of musical space that takes *M* back to *M* (although it may rearrange the points within *M*).

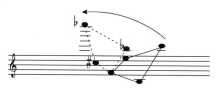

Figure from M. Kugel, 'Translation-Rotation', *Die Reihe* 7 (1965): the arrow indicates an isometry between the two quadrilaterals.

Since a piece of music lasts only a finite amount of time and uses only a finite range of pitches, a musical motif *M* lies within a bounded region of musical space; it has a start and a finish to left and right, and highest and lowest notes above and below. It follows that no translation (except the identity translation that keeps everything exactly where it was) can possibly be a symmetry of *M*. For example, if the translation moves points to the right, it will move the end of the motif to a point of time after the original motif is finished. By the same argument, a glide reflection can never be a symmetry of a musical motif. So only two kinds of symmetry are left: reflections and rotations.

There is no law of mathematics or music to stop composers from using any reflection or rotation they please as a symmetry of their motifs. But in practice, apart from some examples that are too trivial or too abstruse to be interesting, the symmetries of musical motifs turn out to be just two kinds of reflection and one kind of rotation, as follows:

reflection in a vertical line: call this transformation R_v
reflection in a horizontal line: call this R_h
rotation through $180°$ (exactly half a circle): call this R_2

These three kinds of symmetry are not independent; one can prove that if two of them are symmetries of a motif M, then the third is a symmetry of M too. So we can classify motifs according to which symmetries they have, and there are five possibilities, which we shall call the *symmetry types* of musical motifs. (The names *p1* etc. are adapted from crystallography, where one studies the shapes of crystals by describing their symmetries.)

p1: only the identity transformation is a symmetry;
ph: besides the identity, only R_h is a symmetry;
pv: besides the identity, only R_v is a symmetry;
p2: besides the identity, only R_2 is a symmetry;
phv: all of R_h, R_v and R_2 are symmetries.

For some broad guidance, here are examples of letters which (at least approximately) have the five symmetry types. We think of a letter as being a set of points of the plane. For example the reflection R_v is a symmetry of the letter A because the letter covers exactly the same set of points of the plane if you flip it over around a vertical line passing through the top of the A. On the other hand, if you turn A over around a horizontal line, the result is not A but an upside down A, covering a different set of points.

(p1)	F	G	J
(ph)	C	E	K
(pv)	A	M	V
(p2)	N	S	Z
(phv)	H	O	X

The five symmetry types of motifs.

Motifs of the five symmetry types

Type p1: no symmetries

The overwhelming majority of musical motifs belong here. For most of them, this is a fact of no particular significance. But suppose the composer has a mind to use some geometrical transformations such as reflections and rotations. Then an asymmetrical motif M gives much the best value, because its images under the different transformations are all different and distinguishable.

Given that most tunes are highly asymmetrical, what should we say about a composer who takes somebody else's tune, applies a transformation to it and then markets it as his own? This is exactly what Rachmaninov did to a violin caprice of Paganini. Below we show the Paganini original and the inverted Rachmaninov version in lock step. Rachmaninov follows Paganini bar by bar, and it's a chromatic inversion (so that it changes minor to major). But he slightly changes the rhythm and at one point he jumps by an octave. (It's a good exercise to play Paganini's tune upside down with no further alterations; one can see why Rachmaninov made the changes that he did.) What the table doesn't show is the difference in mood between a single violin playing staccato on its E string and a large symphony orchestra souping up the harmonies.

Upper line: Niccolo Paganini, *24 caprices for violin solo*, Op. 1, No. 24.
Lower line: Sergei Rachmaninov, *Rhapsody on a theme by Paganini*, Op. 43, *Variation XVIII*.

In the Classic FM Hall of Fame, where the British public votes for its most popular classical pieces, the scores in the year 2000 were

Rachmaninov inverted version: 33rd out of 300.
Paganini original: nowhere.

Type ph: only reversal of pitch

If a simple melody with no accompaniment has type *ph*, then it consists of a single note repeated. Surprisingly there are motifs with this property. Anton Reicha, a friend of Haydn, published a piano fugue whose subject consists of the same note repeated 34 times. (Towards the end the left hand sees the challenge and manages to repeat a single note 86 times. Fans of the Fibonacci numbers will be interested to hear that these 86 notes are grouped into blocks of 5 beats and 13 beats.)

Reicha's fugue is more entertaining than musical. On the other hand there certainly are worthwhile melodies that lie entirely in one pitch. But then they must owe their interest to another dimension. It could be rhythm, as with drum music, though a good drum player usually varies the timbre as well. The didgeridoo plays only one note, but an expert performer can get a tremendous range of timbres from it. Much of the music of the Italian composer Giacinto Scelsi revolves around changing the timbres of a small number of notes, as in his *Quattro pezzi su una nota sola*.

If a motif of type *ph* has several voices or instruments, the upper voices can reflect the pitch movements of the lower ones and there is no restriction to a constant pitch. When the upper and lower voices move in opposite directions, this is known as *contrary motion*. It forms a well-known gesture in classical music, just as in conversation one sometimes throws one's arms out or brings one's hands together. Some composers do it naturally, as if they never even noticed:

Wolfgang Mozart, *Clarinet quintet*, K381, opening of first movement.

Mozart's two upper voices reflect the movements of the two lower voices very closely for the first nine notes here. The intervals in the upper voices are not always exactly the same as in the corresponding lower ones, as they would be under a mathematical reflection; but they are remarkably close.

Most composers have little interest in making their upper voices mirror their lower ones with mathematical exactness, or in making the contrary motion last more than a few bars. But occasionally a composer does it for interest, as in Bartók's set of 'progressive piano pieces' for people learning to play.

Béla Bartók, *Mikrokosmos, No. 141, Subject and reflection*.

Incidentally, the title of Bartók's piece points to the geometric theme. But mirrors or reflections often appear in the titles of twentieth century compositions:

Boulez, *Constellation-Miroir* (in his 3rd Piano sonata)
Carter, *A mirror on which to dwell*
Debussy, *Reflets dans l'eau*
Francesconi, *String Quartet 3, 'Mirrors'*
Kokkonen, *...durch einem Spiegel...*

Maxwell Davies, *A mirror of whitening light*
Maxwell Davies, *Image, reflection, shadow*
Panufnik, *Reflections*
Ravel, *Miroirs*
Reynolds, *The behavior of mirrors*
Takemitsu, *Rocking mirror daybreak*
Certainly not all of these pieces contain pitch reflections.

Type pv: only reversal of time

One of the earliest recorded pieces of secular music has a strong leaning towards left-right symmetry. This is the jingle recorded on the wall of Reading Abbey, *Sumer is icumen in*.

Anonymous 13th century, *Sumer is icumen in.*

This tune is quite unusual in starting high, dropping and then rising again. A much commoner pattern in the folk music of western Europe (though not in Russia) is to start at a low pitch, rise to a high point and then fall back again. Two typical examples are the French folk song *Nous n'irons plus aux bois* and the Londonderry Air.

First line: French folk song: *Nous n'irons plus aux bois.*
Second line: *Londonderry air.*

The primitive rise–fall pattern is in some sense a grandfather of *sonata form*, where a section of rising tension, starting in the tonic key, is followed by a development section of high tension, and then by a final section where the tension falls and the tonic key is recovered. But in sonata form the final section is never a mirror image of the first section in any more precise sense. In fact musical palindromes, compositions with a virtually exact left-right symmetry, are fairly rare.

Some examples have a programme—these are usually vocal pieces with a libretto—and the symmetry expresses something in the programme. One very effective example of this genre is Stravinsky's depiction of Noah's flood spreading over the world and then receding, in his musical play *The flood*.

Another famous example, with a programme of a sort, has nothing directly to do with reflections and everything to do with balance and stability and the other virtues of a well-ordered state. Here is Handel implying, not quite subliminally, that The Lord God has everything very nicely under control, thank you.

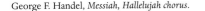

George F. Handel, *Messiah, Hallelujah chorus.*

There is a slight asymmetry in the rhythm, but not enough to damage the symbolism.

Some other musical palindromes seem to have been written for the challenge. If the symmetry is obvious enough, the performers can enjoy it as much as the composer. One of the best specimens of this kind is Haydn's reversible minuet. He was so proud of it that he recycled it into three separate works, the piano sonata shown at the beginning of the chapter, a violin sonata and a symphony.

In a sense, Haydn cheats with this piano sonata. If you take a tape recording of the first half of the minuet and play it backwards, you won't hear anything remotely like the second half. This is because of two physical properties of piano notes. First, they start with a bang and fade out gradually; so if you play them backwards, they start soft and finish with a bang, which makes it impossible to hear them as piano notes at all. And second, the length of time that they last depends on how loud they are (unless the pianist brings down the damper to stop them). This means that a loud note A may start before a soft note B and finish after B; when the tape is reversed, A still sounds as if it was played before B, so the order of the notes is not reversed as it is in Haydn's minuet.

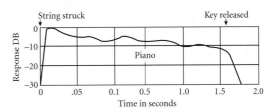

Piano note, from H. F. Olson, *Music, physics and engineering,* Dover, New York (1967), 257.

Luigi Nono's *Canti per tredici* is an exact palindrome for voices and not pianos, but it deliberately uses effects that are like what we hear when we reverse the tape of the piano recording. Like Alban Berg earlier in the twentieth century, Nono used palindrome as a structural device in composition. Most of Berg's and Nono's listeners will not notice these palindromes until they are pointed out, but for composer and performers they bind the music together as a unity.

Type p2: only rotational symmetry

This is not at all a common pattern, and generally it is not easy to hear. It hardly ever happens by accident, except where it falls out of some other feature of the motif.

As with the pattern *pv*, the non-accidental examples tend to be either technical challenges or programmatic symbols. A serious technical challenge should be a complete movement, or two together. This kind of extended symmetry only became possible in twentieth century music; the rules of earlier periods were too rigid. Hindemith provides an example in his piano piece *Ludus tonalis* ('game of tones'). If we ignore the very last chord, the final movement is the same as the first, but rotated through 180 degrees.

—and here follows an hour of music—

Paul Hindemith, *Ludus tonalis*, beginning and end.

Another example is Penderecki's orchestral *Threnody for the victims of Hiroshima*. This is one of many pieces which illustrate the fact that there is no contradiction at all between an emotionally charged topic and a highly formal compositional structure.

To illustrate the symbolic use of *p2*, here is an example with some interesting geometry that its composer may not have been fully aware of.

Nikolai Rimsky-Korsakov, theme from *The golden cockerel*.

The story of Rimsky-Korsakov's opera *The golden cockerel* revolves around a magic bird that sings two songs, one when there is danger and one when there is not. Rimsky-Korsakov has the ingenious idea of making the Safety song a geometric transformation of the Danger song. The tidiest way to do this is to choose a theme that has exactly two images under isometries; so it should be of type *ph*, *pv* or *p2*, not *phv* (which would make it identical under all isometries) or *p1* (which would give it four forms, not two). Should the song be flipped between Danger and Safety by a pitch reflection (as in *ph*) or a time reflection (as in *pv*)? By choosing *p2*, Rimsky-Korsakov gives the answer Yes to both possibilities.

Type phv: all possible symmetries

Interesting motifs of this type are extraordinarily rare. One appears in an elementary piano exercise of Georg Kurtág. The round blobs are instructions to hit the keyboard with the palm of your hand.

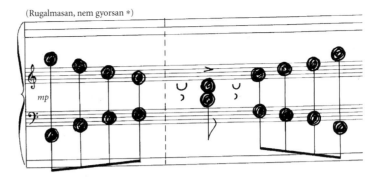

Georg Kurtág, *Játékok for piano I, Hommage à Eőtvős Péter.*

Why are there so few examples of this type? I can only answer with an anecdote. Once I thought I heard an example in a concert of contemporary piano music. Since the composer (Luke Stoneham) was sitting behind me, I asked him in the interval whether I could check the score. When he heard what I was looking for, his jaw dropped and he said that if he had spotted any such figure in the piece, he would certainly have removed it. It seems that any composer with taste regards this symmetry group as too crass to use.

Breaking out of bounds

The dot-dot-dot symbol

• • •

repays some study. We read it from left to right. The second dot comes from the first by a small translation to the right. If we repeat the translation, we get the third dot. That's enough to establish a pattern, and if we made a few more repetitions we would soon run over the edge of the paper.

So the three dots point us to infinity. This is a purely geometrical idea and it transfers immediately to the musical plane. Several composers have used it, usually at the end of a programmatic piece with a message 'and so life goes on'. One such composer is Bedřich Smetana, at the end of his string quartet *From my life* (see overleaf) where the story sinks into the indefinite future.

Béla Bartók used the same device at the end of his opera *Duke Bluebeard's castle*. Shortly before the final 'dot dot dot' we hear (in the German version) the word *ewig* 'ever' repeated four times. One can hear another example (played very softly) at the end of Benjamin Britten's opera *Peter Grimes* when Grimes is gone and the community's life returns to its normal cycle, while 'in ceaseless motion comes and goes the tide . . .'.

Did it have to be a horizontal translation that we used to point to infinity? Yes and no.

This calls for a small digression, to bring in a class of transformations of the plane that includes the isometries and more besides. Defined mathematically, *affine transformations* are the transformations which take any straight line to a straight line. One important kind of affine

Bedřich Smetana, end of string quartet: *From my life*.

transformation is *horizontal dilation* which keeps all pitches the same but slows down the time scale so that notes which were *s* seconds apart become *rs* seconds apart. The number *r* is called the *ratio* of the dilation. If $0 < r < 1$ then the dilation speeds up the time. Likewise a *vertical dilation* expands the pitch scale in some fixed ratio *r* but doesn't alter time.

Vertical dilations occur constantly in Beethoven's writing. He uses them systematically as a way of generating new material out of a basic motif. But the device is much older than Beethoven. There is a kind of canon called a *mensuration canon* where a tune is played simultaneously at two different speeds (and usually at different pitches too). This is a way of using horizontal dilations. It was popular in the fifteenth century, and in the twentieth century several composers used it, most notably Olivier Messiaen (as a metaphor) and Conlon Nancarrow (who used bizarre ratios like

$$\frac{1}{\sqrt[3]{\pi}} \bigg/ \sqrt[3]{\frac{13}{16}} : \frac{1}{\sqrt{\pi}} \bigg/ \sqrt{\frac{2}{3}}$$

in music for a player piano).

To come back to the matter in hand: there are just two kinds of affine transformation of the musical plane that can be iterated as often as we like but eventually lead out towards infinity. These are horizontal translations and horizontal glide reflections, either of which will give us the dot-dot-dot pattern. All other affine transformations of musical space that lead us out towards infinity hit the buffers after a very few iterations: either the pitch rises or falls too far for the instrument, or the music is too quick to be playable, or it's too slow to be heard as music, or some other similar physical problem.

The best composers struggle against these limits, and where necessary they find ways of deceiving the ear into thinking there has been more iteration than in fact there has been. Two examples will suffice.

The first example is Handel fighting against the speed limits built into the action of an eighteenth century organ. The passage is from the organ part of his *Organ concerto in A major*:

George F. Handel, *Organ concerto in A major.*

We think he keeps doubling the speed of the repetition; this is a horizontal dilation with a ratio of 0.5. But when he reaches the physical limit, instead of continuing the iteration by repeating faster, he changes the notes. The ear is deceived. Handel may have learned this or a similar trick from the Italian opera writers.

The second is from one of those sadly beautiful motets that William Byrd wrote for his fellow Catholics (a persecuted minority under Elizabeth I) to sing at Ingatestone House under the protection of Lord Petre, *Non vos relinquam orphanos* 'I will not leave you comfortless'. Jesus is foretelling his ascension into heaven, *Vado* 'I am going'.

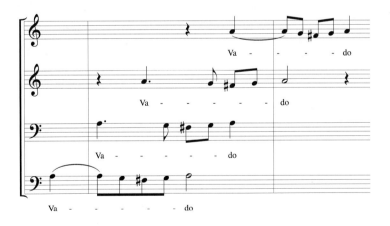

William Byrd, *Non vos relinquam orphanos.*

The moment passes quickly, but this was music to be appreciated by the performers themselves. The *Vado* motif seems to move steadily upward through the voices, pointing to Jesus' own movement upwards to heaven. Geometrically this is a diagonal translation iterated. In fact the movement is not as steady as it sounds; at two of the repetitions there is no movement upwards. Again the ear is deceived.

This passage of Byrd seems to have entered the subconscious of a number of later English choral composers. There is a very similar upward movement in a passage of Gustav Holst's *Hymn of Jesus* to the words 'When I am gone'; and Tippett has a splendid example in the climax of the final chorus of *A child of our time*, to the words 'Walk into heaven'.

Friezes

A *frieze pattern* is a pattern that repeats itself endlessly in one dimension. We can classify frieze patterns by their symmetries, just as we classified motifs. Since a frieze pattern keeps repeating, one of its symmetries must be a translation; this is one difference from motifs. Geometers looked to see what other isometries can be symmetries of a frieze pattern, and they discovered that there are exactly seven symmetry types of frieze. In the chart below, one should imagine each frieze pattern as running infinitely far to the left and the right. The names correspond to those used earlier, except that they also contain *t* for translation or *g* for glide reflection.

(p1t)	FFFFFFFFFFFFFFFFF
(p1g)	FＪFＪFＪFＪFＪFＪ
(pht)	EEEEEEEEEEEEEEE
(pvt)	AAAAAAAAAAAAAA
(pvg)	AＶAＶAＶAＶAＶAＶAＶ
(p2t)	NNNNNNNNNNNNNN
(phvt)	HHHHHHHHHHHHHHH

The seven types of frieze pattern.

A line of music endlessly repeated is a musical frieze pattern. One can find examples of all seven types. But one shouldn't expect too much here; music that repeats itself over and over again is almost by definition background or mood music, not meant to be listened to for its own sake. Nevertheless in at least four cases (*p1t, pvt, pvg* and *phvt*) there are interesting examples.

p1t

This is the type of a pattern that repeats over and over with no symmetries except sheer repetition. Many birds make sounds like this, from the *tap tap tap* of the woodpecker to the *jug jug jug* of the nightingale. Enrique Granados has a famous and suitably repetitive portrait of a nightingale at the end of *Quejas ó la Maja y el Ruiseñor* in his piano suite *Goyescas*. But obviously when birds are mentioned, we have to pay a visit to Olivier Messiaen. Here is slightly less than half of his setting of the song of the curlew. The rest is similar.

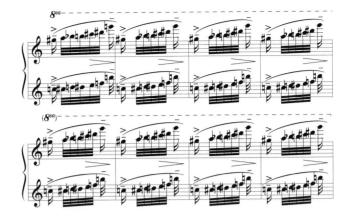

Olivier Messiaen, *Catalogue d'oiseaux, Le courlis cendré.*

pvt

Earlier we saw that the symmetry type of an arch-shape that rises and then falls again is *pv*. So *pvt* is the type of a line of arches; we can see them in the row of As in the example above. Sibelius, that genius of orchestration, gives just such an arch shape to his violins to play over and over again. What makes this an interesting passage is that he does two other things. First, he divides the violins into four groups and makes each group start its arches at a different time. The effect is a throbbing sound that repeats at a quarter of the length of the arch; the arch lasts four bars but the combined pattern repeats at each bar.

Jean Sibelius, *Symphony No. 3*, last movement.

Second, although the violins are providing a background texture, the rest of the orchestra seems not to realise this and keeps trying to turn the rising side of the arch into a foreground tune. It never quite succeeds, but it holds us on the edge of our seats.

phg

A sine curve is the best mathematical example of this frieze type. What music sounds like a sine curve? There are plenty of rippling sounds in music, for example in Smetana's depiction of the Vltava in *Mà Vlast*—though if you look at the score you will see that Smetana's ripples are generally a good deal less regular than the ear takes them to be. But step forward Debussy, whose jumping jacks leap to and fro across the sky in his *Fireworks* prelude. The symmetries are not quite exact, but with music like this, who's counting?

Claude Debussy, *Preludes for piano II, Feux d'artifice.*

phvt

This is the frieze type of a single note repeated regularly and endlessly. The endlessness and the resemblance to a church bell make this figure a potent symbol of death. A repeated note hangs over the last five pieces of Schubert's song cycle *Die schöne Müllerin*, sometimes dryly, sometimes frantically. In the song *Die liebe Farbe* it is constant throughout the piece. In the third bar below, Schubert (always a master of spacing) has placed a huge emptiness between the low D♯ and the high F♯ of the relentless bell. The fact that this is a major chord, which in romantic music tends to express happiness, makes the passage doubly poignant.

Franz Schubert, *Die schöne Müllerin, Die liebe Farbe.*

Three frieze patterns remain. In the following extract, which illustrates *p1g*, the upper staff is the cor anglais, while in the lower staff two bassoons alternately play the frieze motif the right way up and inverted. The inversion is chromatic.

Igor Stravinsky, *The rite of spring*, 14 in score.

The harp motif below illustrates *pht*; one can imagine a row of letter Cs opening up to the left instead of the right. The metric is diatonic in D minor.

Igor Stravinsky, *Petrushka*, 53 in score.

Finally, the flutes and oboes below play a motif that one can see as the crossbar of the repeated N for *p2g* in the table. The motif is made up of whole tones, so that again we are rotating in a chromatic metric.

Claude Debussy, *La mer*, second movement, bar 72.

Conclusion

Books on counterpoint or canon contain many examples of transformations of melodies. Beyond these, one must go to the composers themselves, the recordings and the scores. The following composers are particularly fertile in geometrical ideas:

Johann Sebastian Bach (1685–1750) was the grand master of fugue, and he wrote several collections of fugues which illustrate an amazing range of possibilities.

Béla Bartók (1881–1945) rivalled Beethoven in his ability to spin whole pieces of music out of a few notes by various geometrical transformations.

Ludwig van Beethoven (1770–1827) hardly needs introducing. C. Rosen, *The classical style*, Faber, London (1971), studies Beethoven's development of themes, and compares him in this regard with Haydn and Mozart.

Alban Berg (1885–1935) was, like Anton Webern, a student of Arnold Schoenberg. These three composers developed Schoenberg's twelve-tone techniques, which were built round isometric transformations of a sequence consisting of the twelve notes of a chromatic scale in some fixed order.

Josquin Desprez (c. 1440–1521) was one of a number of polyphonic composers in the period 1350–1500 who built much of their music around types of canon, sometimes of dazzling virtuosity. (Others were Machaut, Dunstable, Du Fay, Ockeghem.)

Joseph Haydn (1732–1809) loved musical tricks and witticisms, but he combined them with a deeply serious commitment.

Olivier Messiaen (1908–92) had an almost obsessive interest in structural devices—for example, scales with particular symmetry types, and rhythmic patterns from classical Indian music. Robert Sherlaw Johnson, *Messiaen*, J. M. Dent and Sons Ltd., London (1989), gives an excellent introduction to Messiaen's methods.

Conlon Nancarrow (1912–98) wrote almost exclusively for player pianos, because these instruments can produce notes with a speed and accuracy which no human player could possibly achieve. His *Studies for player piano* are a kind of modern *Art of fugue*, covering all conceivable kinds of canon. Unfortunately most of them are unpublished and exist only as piano rolls. But recordings have now been issued on CDs and are well worth hearing; the notes issued with the discs are a fascinating introduction.

Campanalogia Improved:

OR, THE

ART *of* RINGING

MADE EASY,

By Plain and Methodical Rules and Directions, whereby the Ingenious Practitioner may, with a little Practice and Care, attain to the Knowledge of Ringing all Manner of *Double*, *Tripple*, and *Quadruple Changes*.

With Variety of *New Peals* upon Five, Six, Seven, Eight, and Nine Bells. As also the Method of calling *Bobs* for any *Peal* of *Tripples* from 168 to 2520 (being the *Half Peal:*) Also for any *Peal* of *Quadruples*, or *Cators* from 324 to 1140.

Never before Publifhed.

The THIRD EDITION, Corrected.

LONDON:

Printed for A. BETTESWORTH and C. HITCH, at the *Red-Lyon*, in *Pater-Nofter-Row*. M.DCC.XXXIII.

CHAPTER 7

Ringing the changes: bells and mathematics

Dermot Roaf and Arthur White

Change-ringers wish to ring bells in different orders, with no bell moving more than one place in successive rows. The mathematical problem is to devise ways of ringing all possible orders (for example, all 5040 permutations of seven bells) without repetition. English bell-ringers solved this problem more than two hundred years ago; about a hundred years later mathematicians began developing the concepts and terminology to tell the ringers that they had been doing 'group theory' and 'ringing the cosets' all along.

When clocks were rare and watches unknown, people needed to be summoned by bells to come to church—large bells make a lot of noise and can call people from long distances. While there may no longer be this time-keeping need, church bells are still in regular use, now often functioning as a musical instrument played by a team, with what turns out to be a strong mathematical aspect to the music.

Bells are hung in bell-towers or belfries and each is sounded by pulling on a rope, one ringer to a bell, which moves the bell and gives energy to its tongue or clapper. A bell makes a nicer noise, and the sound carries further, if it is swung with the mouth upwards rather than while hanging downwards or hit with a hammer. But a bell that swings high swings slowly. The ideal pendulum, swinging at a steady rate regardless of amplitude, is mathematically accurate only for small swings. With larger swings it goes more slowly—indeed, an imaginary bell swinging through a whole circle so that it just reaches the vertical, without going over the top, would take an infinite time to get there. Near the top, a small change in the energy of the bell makes a large difference to the period of a swing, and so another consequence of ringing full circle is that the timing is easy to control.

Each bell sounds near the end of its swing, when the clapper catches up with the bell, sounding once when the bell swings one way, and again as it swings back to its original upside-down position of unstable equilibrium. English ringers control bells with a rope tied to a wheel, fixed to the bell in such a way that they always know which direction

The title page of the third edition of Fabian Stedman's book, *Campanalogia*, published in 1733. Stedman's books were the first to present the underlying ideas of change ringing.

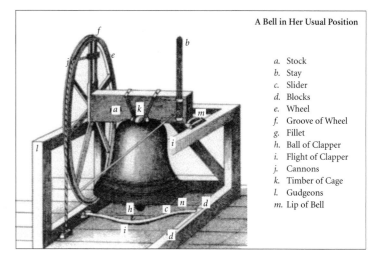

A Bell in Her Usual Position

a. Stock
b. Stay
c. Slider
d. Blocks
e. Wheel
f. Groove of Wheel
g. Fillet
h. Ball of Clapper
i. Flight of Clapper
j. Cannons
k. Timber of Cage
l. Gudgeons
m. Lip of Bell

An early picture showing the details of a bell and its mounting.

the bell is moving. The two different directions are known as *handstroke* and *backstroke*. Each swing is known as a *half-pull* and the complete swing there and back with two sounds is known as a *whole-pull*. (There are smaller hand-held bells to which the mathematical discussion of this chapter also applies, but for simplicity we restrict our attention to bells in belfries.)

There are long gaps between the successive sounds produced by any particular bell—up to two seconds for large bells. Using additional bells helps to fill these gaps, but raises the question of how to combine the sounds produced by the different bells. A variety of choices is possible, though the long interval of time between successive soundings of the same bell always has to be taken into account.

Bells are large and expensive, especially if it is desired for them to be audible at large distances, and only gradually over the centuries has it become common for several separate bells to be installed in the same building; for example, Oxford's first sets of bells, tuned to combine harmoniously, were cast in the seventeenth century.

For up to four or five bells, it is possible to choose the intervals between their pitches so that, even if two or more of them happen to sound at the same time, they produce a harmonious chord; the notes of the full set of bells, rung in order, then form something like an arpeggio. This is the common arrangement in Germany and some other parts of Continental Europe. It is then unnecessary to coordinate the timing of the strokes of the different bells, which are each rung at their own natural speeds, largely determined by their natural weights; in fact, it is this difference in the time interval between strokes on different bells which generates much of the musical interest, by producing a changing variety of melodic sequences, rhythmic patterns and chords.

In the English tradition of full-circle ringing, a different approach has been followed, which allows even larger numbers of bells to be used, without too wide a range of weights.

Normally, the sequence of pitches of the diatonic scale are chosen, producing scales when the bells are rung in order of size—from the lightest bell with the highest pitch, known as the *treble*, to the heaviest bell with the lowest pitch, known as the *tenor*, sounding the keynote. With intervals of pitch as small as tones and semitones, many of the 'chords' produced if two bells strike at once are not harmonious, and it is therefore important to coordinate the timing of the different bells so that they all sound at different times.

Conventionally, the treble is numbered 1, and the other bells are numbered in order of note downwards to the keynote. So, on six bells, the downward scale is 123456. On six bells the last four will be 3456, producing exactly the same melodic sequence as 5678 on eight bells, or 7890 on ten (denoting the tenth bell by 0).

Ringing rounds and call-changes

The simplest (and oldest) procedure was to ring all the bells at the same speed and in regular succession, evenly spaced in what are called *rounds*, with the bells in order from the treble to the tenor; for a set of eight bells, this is:

12345678123456781234567812345678, etc.

A *row* is a sequence in which each bell sounds once; here, each occurrence of 12345678 is a row. To make the rhythm clearer, most teams of ringers leave small gaps after alternate rows: bear in mind that each bell rings successively in opposite directions:

1234567812345678 1234567812345678 1234567812345678, etc.

These gaps occur conventionally after the backstroke row and before the handstroke row.

An evening spent playing unchanging rounds might be considered uneventful, and so the practice developed of changing the order of ringing every so often. Because bells are heavy and slow, this cannot happen rapidly, but two *adjacent* bells can be interchanged without too much difficulty. You might start with rounds 12345678 several times; you might then call out an instruction to the ringers of the second and third bells to exchange the places of their bells in the ringing order, so that 13245678 is rung and repeated until another instruction is called. This way of generating different orders is known as *call-changes*.

Ringers often use the word 'change', both to mean 'row' and to mean the process by which one row is changed to produce another row; here we shall be more precise and use only this latter meaning. Some commonly used phrases can be interpreted in either sense: a 'peal' of 5040 changes' includes 5040 changes as well as 5040 rows, although ringers usually think of the rows as what is being counted; the same applies to 'ringing the changes', the title of this chapter.

Compositions in ringing are designed to include musically attractive sequences, which are usually based on sequences running up or down the scale ('roll-ups'), sometimes with single notes omitted to produce slightly larger intervals of pitch.

Taking every other note of the scale, and running through the scale twice so as to include all the bells, produces 135246 on six bells, or 13572468 on eight. This is the best known of these favourite rows; it is called *Queens*, because a Queen of England is said to have commented on how nice the bells sounded when she heard it being rung.

Reversing the first half of Queens on six bells produces *Whittingtons*: 531246; when heard by Dick Whittington leaving London as 'Turn again, Whittington; Lord May'r of London', it persuaded him to return and, eventually, become Lord Mayor.

Another popular row is *Tittums*, 15263748, in which high and low notes alternate; here the low notes stand out when rung because they are rather louder, producing a shorter but spaced out descending scale.

Seventeenth-century change ringing

Call-changes require a lot of calling on the part of the conductor, the person in charge of a piece of ringing. One way of saving effort for the ringers would be to place a music stand in front of each one showing the successive rows, or to write them on the wall of the belfry, and instruct the ringers as to what new row to ring when the conductor calls 'Change'. This became common in the seventeenth century and still survives in a few remote country towers.

Of course, someone needs to work out the successive rows, making sure that no bell moves more than one place at a time, because more movement would require too much change of speed between one row and the next when a change is made. For four bells this might lead to music like this:

12341234 → 12341234 → 12341234 → 21342134 →
21342134 → 21342134 → 23142314 → 23142314 →
23142314 → 23142314 → 23142314 → 23142314 →
32143214 → 32143214 → 32143214 → 32143214 →
31243124 → 31243124 → 31243124 → 31243124 →
31243124 → 13241324 → 13241324 → 13241324 →
13241324 → 12341234 → 12341234 → 12341234.

Here the conductor has decided to keep the largest bell (4) ringing at a steady speed and change one pair of the other bells every few pulls, not necessarily at regular intervals. (Here, each new row is indicated by bold type.)

This can still sound somewhat monotonous, and as the quality of the bearings on which the bells swing improved, it became possible to change their order more frequently. Indeed, it became common to change after each whole-pull:

$$12341234 \rightarrow 21342134 \rightarrow 23142314 \rightarrow 32143214 \rightarrow$$
$$31243124 \rightarrow 13241324 \rightarrow 12341234$$

—it was then unnecessary for the conductor to call 'Change' each time.

As bearings continued to improve further, the same *touch* (a set of successive rows starting and ending with rounds) could be rung in half-pulls, in between rounds as introduction and conclusion. Here we omit the arrows and use bold type to indicate the pair of bells that have been interchanged from the previous row; for example, 1234 is followed by 2134, which in turn is followed by 2314 and then 3214:

12341234	12341234	12341234	21342314
32143124	13241234	12341234	12341234

This touch was known in the seventeenth century as *ringing the sixes*, because six different rows were rung, each only once, apart from the rounds at the beginning and end. These are in fact all the possible arrangements, and are known as the *extent* on three bells.

After a while, the ringers learnt what to do when ringing the sixes, and no longer needed the rows written down. They would start changing when the conductor called 'Go' and stop in rounds when he called 'That's all'.

Similar methods of ringing different rows on larger numbers of bells were developed. The *twenties* on five bells were:

1234512345	1234512345	2134523145
2341523451	3245134251	3452134512
4351245312	4513245123	5412351423
5124351234	1523412534	1235412345
1234512345	1234512345	

The *twenty-fours* on five bells (with bell number 5 fixed behind) were:

1234512345	1234512345	1234512345
2134523145	2341532415	3214531245
1324513425	3142534125	3421543215
4312541325	1432514235	4123542135
4231524315	2413521435	1243512345
1234512345	1234512345	1234512345

In both twenties and twenty-fours, only one pair is interchanged at a time, so most bells are static. It would sound more interesting to change more than one pair at a time whenever possible. That might mean a bell moving from, say, third place to fourth place and then back to third place in successive rows, which requires larger changes of speed than in the early methods mentioned so far. Such versatility became possible only when better bearings became available.

This practice of *change ringing* started in England and spread throughout the English-speaking world. The first book on the subject was Fabian Stedman's 1668 *Tintinnalogia: or the art of change ringing*, which he followed in 1677 with *Campanologia: or, the art of ringing improved*. The methods developed by Stedman and his fellow-enthusiasts anticipated by a century algebraic ideas that were subsequently rediscovered by mathematicians such as Joseph-Louis Lagrange, in what came to be called the *theory of groups*.

There are now over 5200 sets of bells hung for change ringing in England, another 200 in the rest of the British Isles, and about 100 elsewhere in the world. Of these 5500 sets, 3500 contain 5 or 6 bells, 1700 have eight bells, 200 have ten and 100 have twelve. Recently, two towers have had extra bells added, so that one has 14 and the other 16 bells. A few sets have one or more semitone bells added so as to provide subsets in different keys.

Modern change ringing

The basic idea of modern change ringing is to keep changing the order, and never to ring any particular order more than once, except for rounds which appears at the beginning and end. Each bell may move only one place at a time and, generally speaking, no bell should stay in place for more than two successive rows, except that the tenor may remain at the end all the time. To make learning the changes easier for the ringers, the paths of most bells should follow the same rules, although in most methods one or two bells follow a simpler standard path.

How long does it take to ring every possible row? With three bells there are only 6 rows in the extent, so (at two seconds each) it takes only 12 seconds to ring them all. With four bells there are 24 rows (48 seconds); with five bells there are 120 rows (4 minutes); with six bells 720 rows (25 minutes); and with seven bells 5040 rows (3 hours). That is long enough for most people, so the standard long performance is a three-hour *peal* of 5040 changes. The standard short performance is a *quarter-peal* of 1260 changes, taking approximately 40 minutes, which is typical of the length of time that bells are rung before church services.

When peals are rung on eight or more bells, only a fraction of the full extent can be rung. This allows some scope for selection, so as to include a larger proportion of rows that are preferred for musical reasons; any ringing with 5000 or more different rows is recognised as a peal.

How can we obtain all these different rows?

Three bells

Consider three bells numbered 1, 2, 3 in order from the highest note. The individual bells can move only one place at a time, so there are only two possible ways of ringing the extent, depending on whether the first bell or the last bell holds its place after the first change:

123	132	312	321	231	213	123
123	213	231	321	312	132	123

Four bells

There are several different methods for ringing four bells, one of which, *Plain Bob Minimus* is shown below. (*Minimus* refers to the change ringing of four bells—other numbers of bells have different names, given below.) We first exchange two pairs of bells (denoted by x as all bells move); the next time, we exchange the middle pair, denoted by 14 ('one-four') because the bells in first and fourth places do not move. This alternation is repeated until after eight changes the order would return to rounds if we exchanged the middle pair. We avoid this by changing instead the last pair (denoted by 12). We then repeat all that we have done so far, twice more. We obtain the following pattern.

Note that bell 1 (the *treble*) has a very simple zig-zag path called *plain hunting*. This pattern is similar to that followed by each bell in the case of three bells described above. Next, look at the paths of bells 2, 3

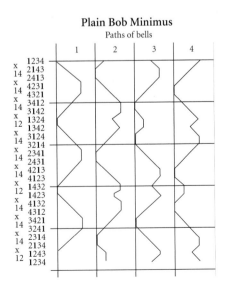

Plain Bob Minimus
Paths of bells

and 4: these three paths are the same, but the bells start at different points—rather like singing a round. These paths are based on plain hunting, with differences every eight changes. If you draw the corresponding paths for the twenty-fours described earlier, you will probably agree that the path of the *working bells* (not including the treble) in *Plain Bob Minimus*—based mainly on plain hunting with steady movement for several changes at a time in each direction—is easier to learn; it is also physically easier to ring, especially on heavier bells. The basic idea is for all the bells to plain hunt until the treble does its two blows leading (at the start of the row): then another bell does two consecutive blows in second place and returns to lead, and the other working bells repeat the positions they have occupied in the previous two rows (known as *dodging*) and continue plain hunting from their new positions until the next time the treble leads.

More bells

This method of ringing works equally well on more than four bells, so the name *Plain Bob* is given to a family of methods—*Plain Bob Doubles* on five bells, *Plain Bob Minor* on six bells, and so on.

Our general principle in choosing methods is that we move as many bells as possible at most of the changes. Now with five bells we can exchange two pairs at each change. We call changes on five bells *doubles*, even though we may occasionally change only one pair—indeed, to obtain all 120 arrangements of five bells, we need to ring at least two single changes. Shortly, we will describe *Plain Bob Doubles*—the Plain Bob method on five bells.

It often sounds better to keep the heaviest bell (the *tenor*) at the end and change only the front bells, as the gap that follows allows the favoured musical combinations at the end of the change to stand out more clearly. As most towers have an even number of bells, this means that an odd number of bells will change their positions. Ringers find steady rhythm and accurate striking rather easier with the tenor 'covering' in this way, and listeners can more easily appreciate the structure of the ringing. Experts, however, find greater challenges and variety in ringing on even numbers of bells, with the tenor continually changing its position.

If seven bells change, we are ringing *Triples*, because there are triple changes—for example, 1234567 changing to 2143657 has three pairs swapping: 12, 34 and 56. The other names for odd numbers changing are *Caters* on nine bells and *Cinques* on eleven bells. With even numbers changing we have *Minimus* (four bells), *Minor* (six bells), *Major* (eight bells), *Royal* (ten bells) and *Maximus* (twelve bells).

With larger numbers, we usually keep the heavier bells working closely together and frequently arrange 'roll-ups', such as ----5678 on eight bells. There are only 24 different roll-ups on eight bells, but composers try to include them all at backstroke since it is felt that the gap that follows improves the music. Other popular sequences are ----6578, ----7568 and ----2468 (as in *Queens*)—all at backstroke.

A peal on eight or more changing bells can include only a small proportion of the possible orders, so the heaviest bells can be kept in fixed relationships to one another. This eases the problems of composition and 'proof' (the verification that rows are not repeated), as well as improving the music.

Some of the music most favoured by composers on more than eight changing bells is related to *Tittums* (see above), which on ten bells is 1627384950 (denoting the tenth bell by 0). Most methods are based more or less closely on plain hunting, and if *Tittums* (or any other change with the heavier bells in the same arrangement) occurs anywhere in a composition, then the same spaced out sequence 7-8-9 and its reverse 9-8-7 with heavy bells alternating with light bells, also occur in several of the preceding and following rows.

In addition—and perhaps valued even more highly—four steps away from *Tittums* in plain hunting on nine bells with the tenth covering (coming at the end), the heavy bells come together at the end of the change to produce the same melodic pattern as the 4-bell equivalent of *Whittingtons*: ------9780. (The 8-bell equivalent of this is ----7568, which was one of the examples quoted above.) This is illustrated by the sequence on the left, which could be produced in *Grandsire Caters*.

Another musical effect, common in all but the most basic methods, and valued both when ringing with and without a tenor behind, is partial repetition, in which one or more pairs of bells dodge (as explained in *Plain Bob Minimus*) while other bells elsewhere in the change must do something different, so that no row is produced twice.) Dodging in the musically conspicuous position at, or just before, the end of the row produces an effect rather like rhymed verse, with rhymes between the ends of alternate lines in groups of four or six. This occurs once every eighteen changes in *Grandsire Caters*, breaking up what is otherwise simple plain hunting; but it occurs continuously, in blocks of six rows, in *Stedman Caters*, which many regard as the most musical of all methods. Here a touch might include the *Tittums*-like rows on the right, in which you can hear the alternation of light and heavy bells.

A non-ringer may like to try to listen to the treble (bell 1), which usually has a path different from (and simpler than) that of the other bells, as in *Plain Bob Minimus* (described earlier) and *Grandsire Caters*. *Stedman Caters* is a rare exception, as here all the bells have the same rather complex path.

Grandsire Caters	*Stedman* Caters
3425196870	9358274610
3241569780	9532847160
2314657980	5938274610
2136475890	5392847160
1263748590	3529481760
1627384950	5324918670
6123748590	5239481760
6217384950	2534918670
2671839450	2359481760
2768193540	3254918670
7286915340	2345196870
7829651430	2431569780
8792564130	4235196870
8975246310	4321569780
9857423610	3425196870
9584732160	3241569780
5948371260	2314657980
5493817620	3216475890
4539186720	3124657980
4351968270	1326475890
3415692870	1234657980
3146529780	2136475890
1364257980	1263748590
1632475890	1627384950
6134257980	6123748590
6312475890	6217384950
3621748590	2613748590
3267184950	2167384950
2376819450	1276839450
2738691540	2178693540

Proof

Recall that one of the ground-rules for change ringing is that the bells should not be rung in the same order more than once, and that checking whether this is the case is called *proof*. For three or four bells it is easy to ascertain that no row is rung twice, by inspection of the written-out rows: simple examination of the extent of *Plain Bob Minimus* shows that the rows are all different. With larger numbers of bells, inspection of all the rows is more laborious. Of course, we can now make computers do the work, but it is reassuring and illuminating to see the proof for ourselves.

Let us look at the 24 rows of *Plain Bob Minimus* and group them in fours.

1234	2143	2413	4231
4321	3412	3142	**1324**
1342	3124	3214	2341
2431	4213	4123	**1432**
1423	4132	4312	3421
3241	2314	2134	**1243**

Here, each set of four can be labelled by the bold row with 1 at the front. If this row is **1abc**, then the others in the set are *a1cb, ac1b* and *cab1*; so, if the rows with 1 at the front are all different, so are the other rows.

There are six rows with 1 at the front:

1234 1324 1342 1432 1423 1243.

These are all different, so the proof is complete. Note that these are just like the six different rows on three bells (123 213 231 321 312 132), with each digit increased by 1 and then 1 placed at the front.

We can now see how to generate the 120 rows on five bells. We start by exchanging the front two pairs (from 12345 → 21435), then the back two (21435 → 24153), and keep repeating. We can group these rows in *leads* of ten rows:

12345 21435 24153 42513 45231 54321 53412 35142 31524 **13254**.

After these nine changes, a double change on the back two pairs would return us to rounds (**13254** → **12345**), so instead we exchange only the pair in the 34 position (**13254** → **13524**). If we then repeat these ten changes three more times, then the rows with 1 at the front become

12345 13254 13524 15342 15432 14523 14253 12435

and the final row is **12345**. So we return to rounds after eight sets of five rows—that is, four leads of ten rows—giving a total of 40 changes; this is the *Plain Course* of *Plain Bob Doubles*.

Let us now look at *Plain Bob Minimus*. The first eight rows are

1234 2143 2413 4231 4321 3412 3142 1324

If we now add one to each digit and put 1 in front, we have the eight rows from Plain Bob Doubles. In order to ring the twenty-four rows of *Plain Bob Minimus*, we make the eighth change from 1324 to 1342, and not from 1324 to 1234.

Now, if we make the 40th change of the Doubles, 12435 → 14235, and then repeat these forty changes, we reach 13425, and after another forty changes we reach 12345; this gives the full extent of 120 different rows. In order that the ringers know what to do, the conductor calls 'Bob' at the 40th, 80th and 120th rows to inform them that the standard pattern has to be altered here.

We can easily extend this method to more bells, with a new change (summoned up by a new call) each time we increase the number of bells. It turns out that more complicated basic patterns enable us to reduce the number of calls to just two.

The problems of proof can involve very sophisticated mathematics. Group-theorists publish learned papers on it, and computer experts produce programs to verify compositions by what group-theorists call 'sledge-hammer' methods.

A simple example is with *Grandsire Doubles*, in which the basic method uses three changes: a double exchange of the front and back pairs (12345 → 21354), a double exchange of the back two pairs (21354 → 23145), and a double exchange of the front two pairs (23145 → 32415). All rows that we can achieve with these changes can be obtained by an even number of single exchanges; these arrangements are called *even permutations* of 12345.

Permutations

It may be helpful to give a brief explanation of permutations. If we take five objects labelled 1, 2, 3, 4, 5, then the ordering 23451 is an *even permutation* of 12345 because it can be obtained by four exchanges of a single pair, whether adjacent or not, and 4 is an even number; for example, we can take:
  12345 → 21345 → 23145 → 23415 → 23451.
Similarly, an *odd permutation* is obtained from an odd number of exchanges; for example, 24351 is an odd permutation of 12345, because it can be obtained by five exchanges:
  12345 → 21345 → 23145 → 23415 → 23451 → 24351.
A permutation cannot be both even and odd. For five bells there are 120 ($= 5 \times 4 \times 3 \times 2 \times 1$) possible permutations in the extent, consisting of sixty even permutations and sixty odd permutations. So, in order to ring the extent, we need to include one single change (such as that from 13254 to 13245).

The 24 rows of *Plain Bob Minimus* fall naturally into three *leads*, in each of which the treble goes from front to back and returns:
  1234 2143 2413 4231 4321 3412 3142 1324
  1342 3124 3214 2341 2431 4213 4123 1432
  1423 4132 4312 3421 3241 2314 2134 1243.
Mathematicians call these twenty-four permutations of four items a *group*, because the permutations satisfy certain relationships when done

successively. The first lead (plain hunting) forms a smaller group, called a *subgroup* of the full group. The three leads are known as *cosets* of the subgroup; together they make up the whole group, and each row belongs to exactly one coset.

Ringers had been ringing leads of *Plain Bob* for many years before mathematicians came and told them that they were actually ringing cosets. But, although ringers had worked out various ways of ringing the extent of 5040 changes of *Grandsire Triples*, they could never do it entirely with pure triple changes (three exchanges of pairs at each change), even though alternate rows are even and odd permutations. Thus, the 'even and odd' argument we used above to show the impossibility of ringing the extent of *Grandsire Doubles* with pure double changes does not apply here.

But ringers went on trying to solve the problem. Eventually a non-ringer, William Henry Thompson, sometime scholar of Gonville and Caius College, Cambridge, was told of the problem and published in 1886 a 17-page pamphlet showing how the leads of *Grandsire Triples* can be grouped together into what are called *Q-sets*. He showed that with pure triple changes the number of *Q*-sets in any composition must be odd. Since an even number of *Q*-sets would need to be rung to complete the extent, the mystery was solved.

After the publication of this work, composers of change ringing started a serious study of group theory; conversely, even non-ringers have found ringing problems worthy of serious study.

A similar problem arose in *Stedman Triples*, where composers also wanted to ring an extent using only triple changes. The calls are called 'bobs', which are triple changes and often rung in pairs (known as *twin-bobs*, which make the ringing easier), but can legitimately be rung separately and 'singles', which are double changes. Many ringers tried for nearly three hundred years to find a composition without singles. Mathematicians equally failed to prove that it was impossible. Recently, however, the niece of one of the authors proved that it is impossible to ring an extent with twin bobs only—and shortly afterwards a ringer composed an extent with odd bobs, but no singles. This composition has since been rung.

Graphs

An important aspect of analysing change ringing and of composing new pieces is to find a mathematical language in which to describe the effect of the changes. We have just seen how the symbolism of permutations and group theory, devised by mathematicians in the nineteenth century, has proved useful in this regard. Another mathematical language, developed in the twentieth century, has a stronger visual quality that simplifies analysis for many people. This involves the use of diagrams

Three bells (singles)

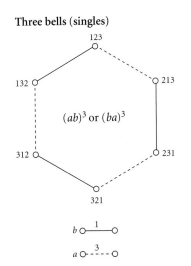

involving points and lines, called *graphs*; note that here we are not concerned with a 'graph' with coordinate axes.

The six rows of *Singles* can be shown as the points of a graph, with lines joining those pairs of points that can be reached by exchanging pairs of bells. If we exchange the front pair, keeping the bell in third place fixed, we use a dashed line and write *a*, while if we exchange the back pair, keeping the bell in first place fixed, we use a solid line and write *b*; for example, 123 and 213 are joined by a dashed line, and 123 and 132 are joined by a solid line. An extent of *Singles* then corresponds to a 'cycle' in the graph, beginning and ending in rounds and visiting each of these six exactly once. The two ways of doing this, clockwise (*ababab*) and anti-clockwise (*bababa*), are readily apparent from the diagram.

The twenty-four rows of *Minimus* can be displayed graphically in a similar manner. Different methods allow different changes: in the *Plain Bob Minimus* we have three changes:

- a simultaneous exchange of two pairs (marked with a solid line);
- an exchange of the middle pair (marked with a dashed line);
- an exchange of the last pair (marked with a crossed line).

The changes are numbered from 1 to 24 (*ababababacababababacababab*).

Four bells (Plain Bob Minimus)

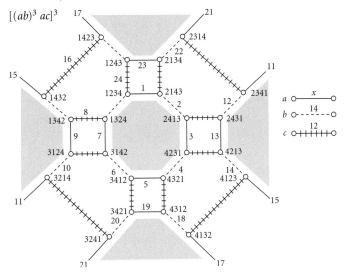

Note that some changes, such as 11 on the diagram, are represented here by two half-lines that must be connected to complete the diagram; that is, the row 3214 (at 8 o'clock) is followed by 2341 (at 2 o'clock), and is then followed by 2431. This *antipodal identification*, when carried out completely, produces a non-orientable surface, called a *projective plane*. This surface can be realised in 4-dimensional space, but cannot be drawn in the 3-dimensional space in which we live, nor in the 2-dimensional space of our diagrams.

Note also that the three-octagon model displays not only the three leads of Plain Bob Minimus, but also the three cosets of the relevant subgroup in the group of all twenty-four permutations.

Most methods have only occasional places where bobs or singles may be called; between these points, the ringing is determined by the initial row. For example, in the *Plain Bob Doubles* described above, the bob can be called only when the treble is at the front—that is, at every tenth change. This means that the treble has the simple plain hunting path: as few bands have enough expert ringers, this is important. So we can describe this method in leads of ten rows:

12345 21435 24153 42513 45231 54321 53412 35142 31524 **12345**

is the first lead which we can summarize as (A) 12345 → 13254;
then, if no call is made, (B) 13524 → 15342;
then (C) 15432 → 14523 and (D) 14253 → 12435;
if now a bob is called, we ring (E) 14235 → 12453;
then (F) 12543 → 15234, (G) 15324 → 13542 and (H) 13452 → 14325;
another bob, and (I) 13425 → 14352;
then (J) 14532 → 15423, (K) 15243 → 12534 and (L) 12354 → 13245;
and the final bob gives 12345.

Five bells (Plain Bob Doubles)

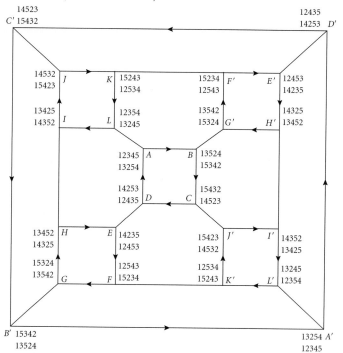

These are shown on the graph; each point represents a lead, marked by its beginning and end; lines with arrows denote the usual route and unmarked lines denote the effect of calling 'Bob'. In this particular case,

Five bells (Plain Bob Doubles on a truncated octahedron)

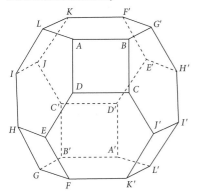

the effect of a bob is the same in both directions. Of course, the leads can be rung forwards or backwards, so each row belongs to two different points, and we must visit exactly one of each such pair of points.

The previous graph was drawn on a plane, but it could equally well be drawn on a truncated octahedron with pairs of leads rung in opposite directions being exactly opposite one another. If we 'identify' each such pair (regarding its two points as a single one), we can extend this antipodal identification to the entire truncated octahedron, and we again obtain the projective plane. Using the above convention of half-lines, we then can represent this graph by another plane figure, but now only involving half as many vertices, edges and regions.

With more bells, we may be calling both *Bobs* and *Singles*, so there will be three different routes from each lead and three routes to each lead.

Modern compositions

We have seen that four types of change are used in *Plain Bob Doubles*: double exchange of the front two pairs, double exchange of the back two pairs, exchange of the bells in positions 3 and 4, and (at the bob) the exchange of the 2–3 pair. The treble is plain hunting and the paths of the other four bells are all alike. The path is symmetrical (the same backwards and forwards); as this makes it easier to learn, composers do their best to create symmetrical methods.

A mathematician might wonder whether all 120 rows can be rung with only three types of change. Suppose that we use only the double exchange of the front two pairs, the double exchange of the back two pairs, and the exchange of the back pair; then the graph has exactly three lines at each point, none with an arrow. The resulting diagram has more than 120 points, because some links between points would so confuse the diagram that it is simpler to represent some rows by two or three points (to some of which only two lines are drawn). There are ten half-lines joining five pairs of rows. The first twenty-four rows (rounds being row 0) of *White's No-Call Doubles* are identified—other identifications can be obtained from the symmetry.

With these identifications around the periphery of the diagram, we obtain the appropriate graph with 120 points (see overleaf). The resulting surface is rather complicated!

The five-fold symmetry of this diagram facilitates the finding of a cycle that leads to an extent. Each set of five points is equivalent under rotation by multiples of 72° about the centre of the diagram. By identifying each such set to a single point, we obtain a simpler diagram with twenty-four points. The surface obtained here by identifying the four half-lines as indicated is called a *Klein bottle*, a one-sided surface which is well known among topologists as a 'bottle' with no inside (so of no use on picnics).

Five bells (White's No-Call Doubles)

$$[(ac)^3 (ab)^3 ac (ab)^2 (ac)^2 ab]^5$$

0	12345	*a*
1	21435	*c*
2	21453	*a*
3	12543	*c*
4	12534	*a*
5	21354	*c*
6	21345	*a*
7	12435	*b*
8	14253	*a*
9	41523	*b*
10	45132	*a*
11	54312	*b*
12	53421	*a*
13	35241	*c*
14	35214	*a*
15	53124	*b*
16	51342	*a*
17	15432	*b*
18	14523	*a*
19	41253	*c*
20	41235	*a*
21	14325	*c*
22	14352	*a*
23	41532	*b*
24	45123	

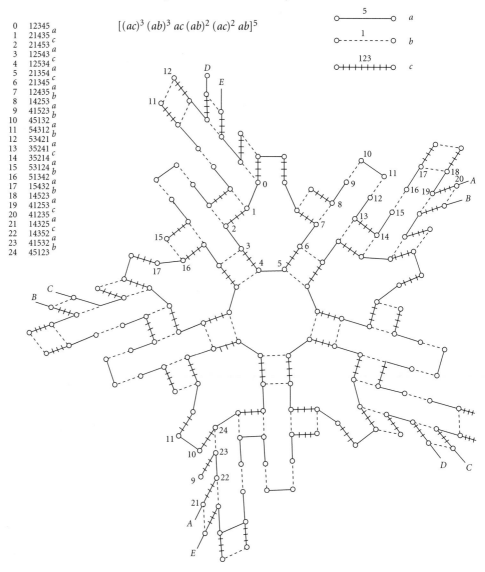

One cycle of length 24 in the simpler diagram, when drawn on the larger graph, ends either at the initial point or at a corresponding point (the initial point rotated through a multiple of 72°). The former case—for example, $(ac)^5 ab(ac)^5 ab$—gives a touch of 24 changes; the latter case—for example, $((ac)^3(ab)^3 ac(ab)^2(ac)^2 ab))^5$—gives the extent of 120 changes. The numbers in the large diagram correspond to the first 24 rows of this extent.

In music thus composed, the plain course is the extent, so no special call (a bob or a single) is required. In December 1984, this method was rung at the Church of St Thomas the Martyr in Oxford, and named by

Five bells (White's No-Call Doubles)

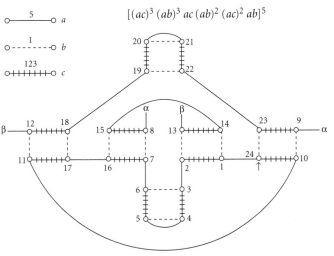

$$[(ac)^3 (ab)^3 ac (ab)^2 (ac)^2 ab]^5$$

the band *White's No-Call Doubles*. This method is not symmetrical, as the backwards version is different, so it was not easy to learn, but exhaustive computer analysis has shown that there is no symmetrical three-change Doubles method that has the extent as its plain course.

A variant of this method is known as *Reverse White's No-Call Doubles*, and was rung at the Carfax Tower, Oxford, in February 1985. This basically turned the previous piece back to front, by using the double exchange of the back two pairs, the double exchange of the front two pairs, and the single exchange of the front pair—instead of the double exchange of the front two pairs, the double exchange of the back two pairs and the exchange of the back pair. A similar diagram led to another piece, *Western Michigan University Doubles*, which was first rung at the Carfax Tower in July 1987.

Nor is the appeal of change ringing today confined to the United Kingdom. In April 1991, the first Irving S. Gilmore International Keyboard Festival was held in Kalamazoo, Michigan, USA. The programme on the opening night featured the world première performance of Kalamazoo composer C. Curtis-Smith's *Concerto for left hand and orchestra*, written for and performed by Leon Fleisher. The final movement of this concerto incorporates elements from change ringing, including *Plain Bob* and *Western Michigan University Doubles*, to great effect in both piano and orchestra. This was a noteworthy occasion for the introduction of change ringing, through the programme notes and the music itself, to an American audience of nearly four thousand people.

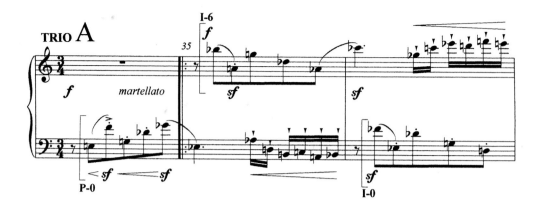

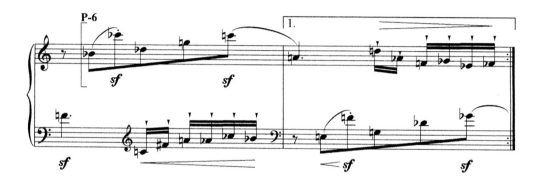

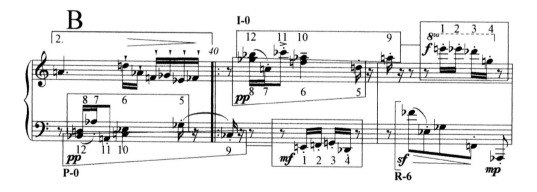

CHAPTER 8

Composing with numbers: sets, rows and magic squares

Jonathan Cross

Throughout the twentieth century mathematical ideas emerged as basic tools for the composer. Here we consider a range of these, from the twelve-tone row of Arnold Schoenberg and the magic squares of Peter Maxwell Davies to the use of set theory and geodesic surfaces by Iannis Xenakis.

Accusations of lack of artistry, lack of creative imagination, and even lack of musicality have been hurled by critics and music-lovers alike at very many twentieth-century composers, and not least at the Viennese composer, Arnold Schoenberg. His 'discovery', as he put it, in the early 1920s, of his 'method of composing with twelve tones', was seen by those at a distance from his work as being a kind of compositional equivalent of those 'painting by numbers' kits that can be bought in children's toy shops. In Schoenberg's composing kit was to be found the composer's equivalent of paint and brushes—namely, the twelve notes of the chromatic scale, arranged in any order of the composer's choosing, so long as each note appeared only once (a 'tone row'). The over-printed canvas—for Schoenberg—was often a ready-made form from the musical past: a movement from a Baroque suite, a waltz, or even a sonata form movement. Onto this canvas the tone row was laid, according to very straightforward mathematical operations of translation (or, in musical terms, transposition) and different kinds of mirroring (inversion and retrogradation).

Put in this way, it is hardly surprising that Schoenberg has recurrently been misrepresented as the bogeyman of twentieth-century music. Nineteenth-century Romantic thought had led us to believe that the composer was someone special, almost a god, set apart from the rest of ordinary society. He was someone in touch with the muses who waited for inspiration to strike before pouring out his soul, by means of some mystical process, in order to produce works of art to be revered by the masses almost as if they were holy relics—a surprising attitude, one might think, for an age which paradoxically saw the rapid development of logical scientific knowledge and method. Schoenberg *seemed* to be

The opening of the Trio from the Minuet and Trio of Schoenberg's *Piano suite*, Op. 25, showing the disposition of the six forms of the basic tone row used in its composition.

suggesting quite the opposite. Twelve-note music had nothing to do with inspiration, or even with musicality, but was seen as mechanical or, worse still, mathematical.

However, this is a misguided view. As will be seen below, many composers of the twentieth century found numbers and various mathematical models a useful source of compositional material or of processing material. In the hands of some, the results certainly are mundane and mechanical. But mundane and mechanical music is possible under any system—not least, tonality. It is the creative use to which such number systems are put that makes for a 'successful' piece of music, not the fact that numbers in themselves have been compositionally deployed.

Why, then, did Schoenberg feel it necessary to invent the twelve-note method? The answer to this question should tell us much, not only about Schoenberg's peculiar historical predicament, but also about why so many composers in recent decades have attempted to frame their music within the context of mathematics.

Arnold Schoenberg

Schoenberg presided over the break-up of tonality, the system that had governed the composition of music for 300 years. When, in 1907, he finally abandoned a key-signature in the finale of his *2nd string quartet*, it was not a wilful attempt to destroy the past; rather, it was an inevitable and necessary step. Tonality had reached the end of its useful life; it could no longer contain the extreme levels of chromaticism and dissonance that had developed in music. The dissonance had to be emancipated.

But with the abandonment of tonality, Schoenberg was confronted with the problem that nearly all composers of the twentieth century had to face. Where was he now to begin? There was no obvious context, no common practice within which to start writing. With every piece he had to begin afresh, had to create his own rules and modes of operation, his own structures. At first he was able to write only very short or fragmentary pieces, or was compelled to rely on texts to structure the music. But eventually he moved to a position where he began to use contrapuntal techniques to provide a more *logical* structure, and eventually this became codified in the twelve-note system. His aim in adopting the 'method' was to provide *comprehensibility* (out of the 'chaos' of free atonality), its main advantage, he claimed, being its unifying effect: 'In music there is no form without logic, there is no logic without unity'. The rigour, the mathematical logic, of the twelve-note system was, in some senses, a substitute for the logical rules of the tonal system.

However, and this is perhaps the most important thing, Schoenberg did not see the method as a general panacea for the ills of twentieth-century music. Far from it: 'The introduction of my method of composing with twelve tones does *not* facilitate composing'. The method merely provided a logical context within which composition could take place. As he wrote, 'One has to follow the basic set; but, nevertheless, one composes as before', a view echoed almost exactly by his pupil Webern, 'For the rest, one composes as before, but on the basis of the row'.

Composers of the twentieth century found many ways around this central problem. Some adopted and adapted Schoenberg's method; others, as we shall see, drew on mathematical sources such as set theory, game theory, magic squares, Fibonacci numbers, and so on, to provide them with material or methods of working. Neither the method, nor mathematics, nor any other system, has made the actual act of composition any easier, nor (necessarily) any more mechanical.

Let us consider an example of Schoenberg's twelve-note practice: the Trio from the Minuet and Trio of the *Piano suite*, Op. 25 (1921–3). Figure 1 shows all the material for the Trio. The form of each row is indicated by a letter: *P* = prime (or original), *I* = inversion, *R* = retrograde (the prime form backwards) and *RI* = retrograde inversion (inversion backwards). Of the 48 possible forms of his twelve-note row, Schönberg uses just six in the Trio: the row itself (*P*-0), the row transposed up a tritone (*P*-6), this transposition backwards (*R*-6), the inversion of the row (*I*-0) and its tritonal transposition and retrograde (*I*-6 and *RI*-6).

Already Schoenberg is having fun with the peculiar properties of this row, with certain patterns that remain constant across the geometric transformations. For instance, the row spans a tritone from E (note 1) to B♭ (note 12), so that by employing transpositions only of a tritone, each form of the row will begin with either an E or a B♭. This interval will then become explicitly represented in the music when, as happens in the

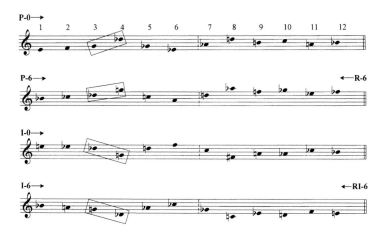

Figure 1. The six forms of the row used in the Trio.

Trio, the six forms of the row are strung together. Another invariant across the transformations is a further tritonal pair G–D♭ (symmetrically placed within the E–B♭ pair), and a feature is made of this in the music.

So how does this manifest itself compositionally? The extract at the beginning of the chapter shows us. What we see and hear is a beautifully formed piece of geometry realised in music. The Trio is canonic throughout and the structure of the row and its transformations articulates these canons clearly. In the first half we have a canon at the tritone in inversion at a bar's distance: *P-0* imitated by *I-6*, and *I-0* imitated by *P-6*. The second half splits the row into groups of four, still an inversional canon but now at the octave (*P-0* imitated by *I-0*). Finally, we return to two-part counterpoint where *R-6* is imitated by *RI-6*.

Thus, though Schoenberg has followed the basic set throughout, nevertheless, in terms of transpositions and deployment, and in terms of rhythms, registers, dynamics and form, Schoenberg has composed freely. The row provides intervallic material; it does not do the composer's work for him.

Interestingly, Schoenberg was not the first to invent a twelve-note system—such ideas were evidently 'in the air' in Vienna in the earlier years of the century. Josef Matthias Hauer, a Viennese contemporary of Schoenberg, had already devised a different system of composing with all twelve notes before Schoenberg. Hauer's ideas were based on what he described as cosmic laws, and (notably) he proposed that music—specifically, atonal music—represented a supreme kind of mathematics.

Alban Berg

Schoenberg's pupils quickly followed their teacher's example by adopting the twelve-note method. The first substantial work of Alban Berg's to use the method (although not in every movement) is the *Lyric suite* for string quartet of 1926. The outer sections of the 3rd movement, the 'allegro misterioso', employ the method: indeed, its structure is dependent on a simple mirroring device where two-thirds of the first 69 bars of the movement are mirrored *exactly* in the last 46 bars and frame a central, more freely atonal section of 23 bars (see Figure 2).

These numbers are highly significant because there is another sense in which Berg was 'composing with numbers' in the *Lyric suite*, and this has to do with its proportional relations: both the durational lengths and the tempi of movements. There are various numerological clues in the score, but the extent to which number symbolism, as well as other kinds of cryptograms and enigmatic quotations, govern the structure of the work, was first fully revealed by George Perle in 1977. Berg, it seems, was obsessed with the number 23. It evidently had some great

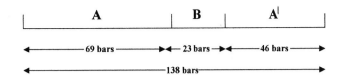

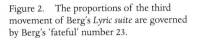

Figure 2.   The proportions of the third movement of Berg's *Lyric suite* are governed by Berg's 'fateful' number 23.

personal significance for him—he referred to it as his 'fateful' number. If we look again at the 'allegro misterioso', we can see that its proportions, in terms of numbers of bars, are governed by multiples of 23. This is no fluke. Movements 1 and 4 are both 69 bars long (3×23); movement 5 is 460 bars long (20×23); and movement 6 is 46 bars long (2×23). Furthermore, the metronome markings are also multiples of 23: movement 4, $\downarrow = 69$; movement 6, $\downarrow = 69$ and $\downarrow = 46$.

As for the length and tempi of the other movements, they are all multiples of 10: metronome marks of 100 or 150 and a second movement that is 150 bars long. Note that the length of movement 5 (460 bars) is a multiple of both 23 and 10. What is the significance of this 10? Some detective work by Perle, including a reading of Berg's letters and the discovery of a miniature score meticulously and colourfully annotated by Berg, revealed that 10 was the fateful number of Mrs. Hanna Fuchs-Robettin with whom, it transpires, Berg had become passionately involved. The score is secretly dedicated to her in Berg's own hand:

It has also, my Hanna, allowed me other freedoms! For example, that of secretly inserting our initials, HF and AB, into the music, and of relating every movement and every section of every movement to our numbers, 10 and 23. I have written these, and much that has other meanings, into the score for you ... May it be a small monument to a great love.

Thus the intertwining of 10 and 23 has not only structural implications for the composer but strongly extra-musical (extra-marital?) ones too. It remains a fascinating personal example of composing with numbers.

Anton Webern

The late works of Anton Webern, Schoenberg's other celebrated pupil, are concise statements and show a highly developed understanding of the possibilities of the twelve-note method, particularly in terms of their concentrated motivic working and their exploration of symmetrical structures. Canons abound. Yet the end results are not in any sense mechanical or abstractly mathematical but poignantly expressive. As one commentator has observed about Webern's serial string quartet: 'its "suitability for study", as a compendium of Webern's serial technique in full maturity, should not blind us to its *musical* qualities'.

The twelve-note row with which Webern composed his *Concerto*, Op. 24 (1934), is given in Figure 3a. It is a marvellous example of symmetry, even within the row itself. Each half of the row involves a

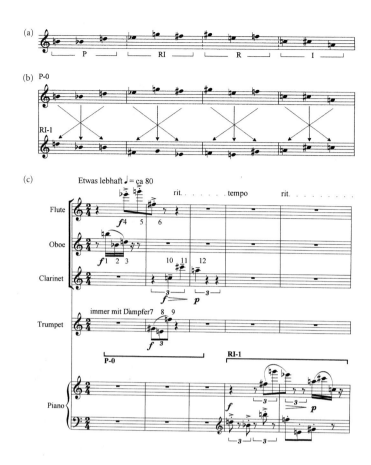

Figure 3. Webern's *Concerto*, op. 24.
(a) The basic row, showing its four three-note subsets.
(b) The *P*-0 and *RI*-1 forms of the row, showing the identical pitch-class content of each three-note set.
(c) The opening shows the use of the *P*-0 and *RI*-1 forms of the row divided into three-note groups, each containing a semitone and a major third.

mirror symmetry and the row can be further broken down into four groups of three notes, each of which contains the intervals of a semitone and a major third, and which represent, in microcosm, the four different forms of the basic row—prime or original, retrograde inversion, retrograde and inversion. Furthermore, the retrograde inversion form of the complete row in transposition (a semitone) preserves the pitch-class content of the three-note groups (see Figure 3b). Figure 3c, the opening of the first movement, shows how this is exploited in the actual music. Notice how Webern makes a rhythmic feature of the three-note groups.

Thus, numbers again provide a context within which the composer can work; they are in no sense the end result—that is, what the piece is about—which is more than can be said for the ways in which Webern was interpreted by some of the younger generation of avant-garde composers after the Second World War. The works of Webern, not Schoenberg, were viewed as the models for the future of music. Only total organisation of music in all its aspects (pitch, duration, mode of attack, dynamics, form) meant that the composer, in theory, was in complete control of the music and independent of forms and processes from the past. Olivier Messiaen was one of the first to suggest the possibilities of total serialism in his *Modes de valeurs et d'intensités* for piano of 1949, but it was his pupil, Pierre Boulez, who took these ideas to their extreme and logical conclusion in his *Structures* for two pianos of 1952. Karlheinz Stockhausen and, on the other side of the Atlantic, Milton Babbitt were similarly extending serial principles beyond the domain of pitch.

Pierre Boulez

The very title of Boulez's *Structures* gives away its central premise—namely, that it is concerned with building integrated musical structures that stand on their own terms rather than being dependent on anything outside of themselves. The architectural implications of the title were intentional and exemplify a more general trend (and not just in music) towards associating art with science, mathematics and architecture. The development of the possibilities of electronics in music was just one reason for this—and the concomitant scientific exploration of the properties of sound. Varèse anticipated this in works with such titles as *Density 21.5, Ionisation* and *Hyperprism*. Later composers made explicit use of these ideas in works with such titles as Cage's *First construction (in metal)*, Boulez's *Polyphonie X* and Stockhausen's *Zeitmasse*. All these works represented a desire on the part of the composers to move forward, to eradicate the past and memories of earlier music; the apparent 'objectivity' of number, mathematics and the mathematically based architecture of a figure like Le Corbusier provided a means to achieve this.

The structure of Boulez's *Structures* is based entirely on the basic row from Messiaen's *Mode de valeurs*—see Figure 4a. Two number matrices were derived from this to represent all 48 forms of the row which are used once each in *Structure Ia*. Each pitch class corresponds to the same integer throughout: $E\flat = 1$, $D = 2$, $A = 3$, etc.

From these matrices a series was also derived for durations by reading each integer as numbers of demi-semiquavers. For example, at the very beginning Piano I plays the pitch classes of the original row, but with the durations of the final inverted, retrograded row (12, 11, 9, 10, 3, . . .)—see Figure 4b.

Furthermore, each statement of the row was assigned a particular dynamic and mode of attack determined by the matrices—Figure 4c shows the row of 12 dynamics and 10 modes of attack. The selection of dynamic and mode of attack is determined by reading diagonally across the matrices: the **P**-matrix for Piano I, the **I**-matrix for Piano II. Even the order in which the 48-note and 48-duration series are chosen is determined by the number matrices: for instance, the first twelve-note series in Piano I are those of the **P**-matrix but in the order of the numbers of the first row of the **I**-matrix (1, 7, 3, 10, . . .).

(a) (i) Prime or original form of row

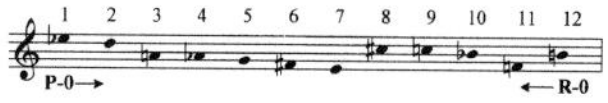

(ii) Inverted form of row

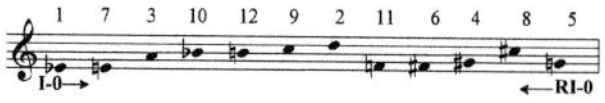

1	2	3	4	5	6	7	8	9	10	11	12
2	8	4	5	6	11	1	9	12	3	7	10
3	4	1	2	8	9	10	5	6	7	12	11
4	5	2	8	9	12	3	6	11	1	10	7
5	6	8	9	12	10	4	11	7	2	3	1
6	11	9	12	10	3	5	7	1	8	4	2
7	1	10	3	4	5	11	2	8	12	6	9
8	9	5	6	11	7	2	12	10	4	1	3
9	12	6	11	7	1	8	10	3	5	2	4
10	3	7	1	2	8	12	4	5	11	9	6
11	7	12	10	3	4	6	1	2	9	5	8
12	10	11	7	1	2	9	3	4	6	8	5

'P' MATRIX

1	7	3	10	12	9	2	11	6	4	8	5
7	11	10	12	9	8	1	6	5	3	2	4
3	10	1	7	11	6	4	12	9	2	5	8
10	12	7	11	6	5	3	9	8	1	4	2
12	9	11	6	5	4	10	8	2	7	3	1
9	8	6	5	4	3	12	2	1	11	10	7
2	1	4	3	10	12	8	7	11	5	9	6
11	6	12	9	8	2	7	5	4	10	1	3
6	5	9	8	2	1	11	4	3	12	7	10
4	3	2	1	7	11	5	10	12	8	6	9
8	2	5	4	3	10	9	1	7	6	12	11
5	4	8	2	1	7	6	3	10	9	11	12

'I' MATRIX

(b)

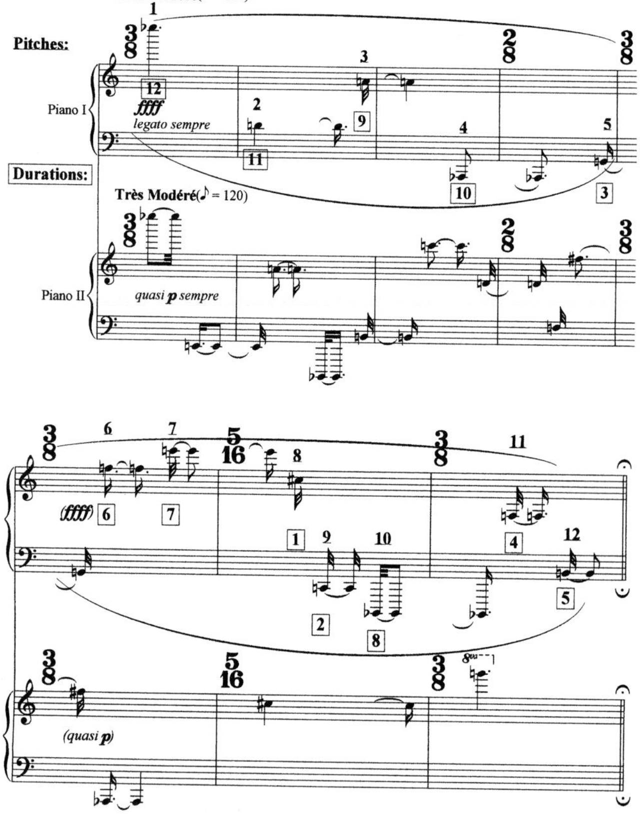

(c) DYNAMICS ROW

1 2 3 4 5 6 7 8 9 10 11 12

pppp ppp pp p quasi p mp mf quasi f f ff fff ffff

MODE OF ATTACK ROW

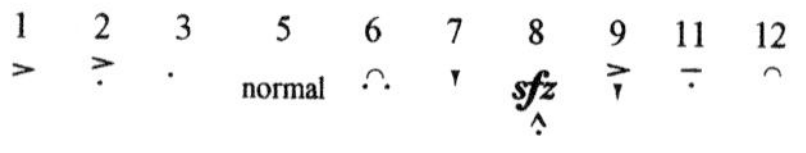

Figure 4. Boulez, *Structure Ia*.

Where does this leave the composer? What scope is there for him or her, in Schoenberg's words, to compose 'as freely as before'? Not too much, apparently. Though Boulez makes free choices regarding register, tempo, metre, and even the use of rests, his hands were tied by the system. The end result is so highly over-determined that it ends up sounding almost completely random: the differences are not readily discernible between *Structure Ia* and, say, Cage's near-contemporary *Music of changes*, where chance procedures of coin-tossing and use of the I-Ching were used to determine the various musical parameters. As an experiment in number made audible, Boulez's *Structures* are fascinating, but he was soon to admit that 'composition and organisation cannot be confused with falling into a maniacal inanity, undreamt of by Webern himself'. Whether or not *Structures* is maniacally inane is for the individual listener to decide.

Peter Maxwell Davies

The English composer, Peter Maxwell Davies, began his composing life as a follower of the thinking of Schoenberg and showed an early familiarity with serially derived techniques of composition. There is, as Paul Griffiths has pointed out, a kinship between the work of Maxwell Davies and Boulez of the mid-1950s 'in matters of rhythmic style, texture and serial handling'. Though their paths have subsequently gone in very different ways, there is a striking similarity in their attitude to number in generating musical material in some of their works. In particular, procedures in those works of Maxwell Davies of the 1970s which 'process' pitch and durational material through magic squares are not that dissimilar from some of Boulez's working in *Structures*.

Ave maris stella (1975) is one such example in which the Gregorian chant 'Ave Maris Stella' is, in Maxwell Davies's words, ' "projected" through the magic square of the moon'. *A mirror of whitening light* (1976–7) is another. The title, according to the composer, refers to the alchemical process of purification or 'whitening', 'by which a base metal may be transformed into gold, and, by extension, to the purification of the human soul'. The 'agent' of this transformation is the spirit Mercury, represented here by the magic square of Mercury, and through which is projected the plainchant *Veni sancte spiritus*. The number 8, Davies tells us, 'governs the whole structure', and its source is the 8 × 8 'Magic square of Mercury' in Figure 5a, in which each row and column and each diagonal adds up to 260.

Figure 5b shows the way in which the plainchant is projected through the magic square. An 8-note 'summary' was derived from the beginning of the chant and consists of 8 different pitches, though it still maintains the profile of the original. An 8 × 8 matrix was then constructed in which each note of the summary was transposed, just like a tone row, to begin on each of its constituent notes, and each note was numbered consecutively from 1 to 64. The final stage was to map this matrix on to the magic square.

The composer then charted various courses through the matrix to generate pitch material: from top to bottom, left to right; from bottom to top, right to left; diagonally; in spirals; indeed, in any way he chose. Figure 5c shows how this is achieved at the opening of the work—in this case, top-bottom, left-right (C, A, B♭, F♯, D, D, . . .). It is just one way in which notes were generated in this piece, one aspect of the various transformations or 'whitenings' that the plainchant undergoes.

Durational lengths can also be determined by the Mercury matrix—Figure 5d shows one such instance. The pitches of the clarinet line were generated by starting at 'square 47' (see Figure 5b) and working backwards and upwards:

47 [B♭], 17 [F], 33 [F♯], 31 [D♯], 30 [E], . . .

The durations of the accompanying bassoon line use the same numerical sequence from the magic square, but here all the numbers were converted so that they lie within the range 1 to 8, by reducing them modulo 8 (for an explanation of modular arithmetic, see Chapter 9). This new but related numerical array was then taken to represent numbers of quaver beats and is stated in the opposite direction from the pitch 'row':

clarinet 'pitch row'	47	17	33	31	30	36	37	27	26	40
bassoon 'duration row'	7	1	1	7	6	4	5	3	2	8

The bassoon's pitches, incidentally, were generated by a left-right reading of the Mercury matrix starting, as at the beginning of the work, in the top-left corner.

Can any of this be heard? Maxwell Davies has great faith in his listeners: these 'sequences of pitches and rhythmic lengths . . . [are] easily memorable once the "key" to the square has been found', he claims. No doubt he would argue that the 'logic' given to the various transformations by the magic square is, at the very least, subconsciously perceived. I have my doubts. What one *hears* is a piece of music, clearly structured with a focal 'key' centre of C, and not just a mathematical game made audible. However, the numbers were vital to the compositional process, as they were a means of providing the composer with his working material. To misappropriate Schoenberg, one has to follow the magic square; but, nevertheless, one composes as before. As Maxwell Davies himself has said in the context of his later *Second symphony*, magic squares 'are a gift to composers if used very simply as an architectural module'.

(a) Magic square of Mercury

8	58	59	5	4	62	63	1
49	15	14	52	53	11	10	56
41	23	22	44	45	19	18	48
32	34	35	29	28	38	39	25
40	26	27	37	36	30	31	33
17	47	46	20	21	43	42	24
9	55	54	12	13	51	50	16
64	2	3	61	60	6	7	57

(b)

ve - ni Sanc - te Spi – ri – tus re - ple tu - o - rum cor - de fi - de - li - um

Summary (derived from the plainchart and retaining its profile)

8 × 8 matrix

G 1	E 2	F 3	D 4	F# 5	A 6	G# 7	C 8
E 9	C# 10	D 11	B 12	D# 13	F# 14	F 15	A 16
F 17	D 18	E♭ 19	C 20	E 21	G 22	F# 23	A# 24
D 25	B 26	C 27	A 28	C# 29	E 30	D# 31	G 32
F# 33	D# 34	E 35	D♭ 36	F 37	A♭ 38	G 39	B 40
A 41	F# 42	G 43	E 44	G# 45	B 46	B♭ 47	D 48
G# 49	F 50	F# 51	E♭ 52	G 53	B♭ 54	A 55	C# 56
C 57	A 58	B♭ 59	G 60	B 61	D 62	C# 63	F 64

Magic square of Mercury

C 8	A 58	B♭ 59	F# 5	D 4	D 62	C# 63	G 1
G# 49	F 15	F# 14	E♭ 52	G 53	D 17	C# 10	C# 56
A 47	F# 23	G 22	E 44	(D) G# 45	E♭ 19	D 18	(G#) D 48
G 32	D# 34	E 35	C# 29	(D) A 28	A♭ 38	G 39	(A) D 25
B 40	B 26	C 27	F 37	D♭ 36	E 30	D# 31	F# 33
F 17	B♭ 47	B 46	C 20	E 21	G 43	F# 42	A# 24
E 9	A 55	B♭ 54	B 12	D# 13	F# 51	F 50	A 16
F 64	E 2	F 3	B 61	G 60	A 6	G# 7	C 57

Figure 5. *Peter Maxwell Davies, A mirror of whitening light.*

Figure 5.   Peter Maxwell Davies, *A mirror of whitening light*.

Figure 5. Peter Maxwell Davies, *A mirror of whitening light*.

Figure 6.   Xenakis, *Metastasis*.

Iannis Xenakis

Such a sentiment was also close to the heart of another composer for whom an understanding of mathematics and architecture were fundamental. Iannis Xenakis was born of Greek parentage and educated in Greece; ancient Greek culture—be it drama, architecture, philosophy or mathematics—continued to have a profound influence on his thought.

Xenakis's early education was principally as an engineer, and when he moved to Paris in 1947 he not only studied with the composers Honegger, Milhaud and Messiaen, but also met the architect Le Corbusier with whom he was to collaborate on a number of important projects. Their most celebrated undertaking was for the Philips Pavilion (Figure 7) at the 1958 Brussels World Fair for which, in just a few days, Xenakis sketched the basic structure using conoids and hyperbolic paraboloids. As Xenakis later observed:

I discovered on coming into contact with Le Corbusier that the problems of architecture, as he formulated them, were the same as I encountered in music.

And elsewhere:

With Le Corbusier I discovered architecture; being an engineer I could do calculations as well, so I was doing both. This is quite rare in the domain of architecture and music. Everything started coming together and I also asked musical and philosophical questions.

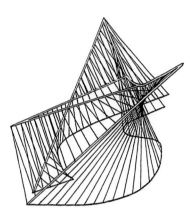

Figure 7.   First model of the Philips pavilion. Its structure is generated by straight lines.

It would seem, then, that for Xenakis music and architecture were concerned with the same issues: in architecture his ideas were articulated in space; in music they were articulated in time. Furthermore, mathematical models underpinned the development of his ideas in both realms.

His first acknowledged composition, *Metastasis* ('transformations') of 1953–4, clearly exemplifies these concerns. The structure of the curved surfaces of the Philips Pavilion was generated by straight lines; *Metastasis* had already demonstrated, as Xenakis put it, that it was 'possible to produce ruled surfaces by drawing the glissandi as straight lines'. Music and architecture here found an intimate connection, as we can see if we compare Xenakis's graph plotting the paths of a section of glissandi with the same passage in the score—see Figure 6.

Metastasis shows Xenakis exploring architecturally derived notions of mass and ruled surface, and a concern to represent 'sound events made out of a large number of individual sounds [which] are not separately perceptible,...[to] reunite them again...[so that] a new sound is formed which may be perceived in its entirety'. In *Metastasis* one is not aware of individual sounds but of a new mass of sounds and timbres. The means by which he achieved this were derived from *The modulor* of Le Corbusier: pitches (based on twelve-note rows) were assigned a series of durations based on the Fibonacci sequence, along with a range of timbres. The way in which this material was processed *became* the form of the piece.

Xenakis subsequently developed these ideas in a much more comprehensive way, using many mathematical models as well as computers to assist him in his pre-compositional calculations. He soon became interested in probability theory as a way of handling mass sound phenomena, and from this grew what he described as 'large number' or 'stochastic' music, where the operation of individual elements is unpredictable but the shape of the whole can be determined. For example, *Pithoprakta*, the next work after *Metastasis*, drew (the composer claimed) on Maxwell–Boltzmann's kinetic theory of gases; *Achorripsis* employed Poisson's law; and *Duel* and *Stratégie* used game theory—each work employed two conductors who 'compete' with one another. More recently, Xenakis developed what he called 'symbolic music' which drew on principles of symbolic logic. Paul Griffiths has observed that 'Xenakis's symbolic music has... the nature of a translation into sound of theorems of set theory', first evident in *Herma* for piano of 1960–1.

This may suggest that Xenakis's music is completely abstract and sterile. Not at all. His music, like the man, is all too human and he always asserted the primacy of music over mathematics—music, he believed, is never reducible to mathematics, even though they have many elements in common. Xenakis was a philosopher who expressed his ideas primarily in music, but who was constantly searching for profound fundamental principles that underlie all thought. As another commentator has put it, 'he gives us something only an artist can give—a dynamic picture of the universe informed by the science of today'.

Although Xenakis's use of a variety of mathematical models may have been undertaken in a more consistent and thoroughgoing manner than almost any other composer, it does not make his music any less exciting, challenging, creative—or even valid—than music composed in a different age or by different means. Mathematics is a means to an end, not the end in itself. Composers today are as aware as have been thinkers of the past that music is inherently mathematical, but this does not mean to say that it *is* mathematics. Composing with numbers is not an admission of compositional failure, a substitute for 'inspiration' or 'musicality', whatever those concepts may mean. Composers have composed with numbers as one way of generating new musical ideas, as a means of stimulating their creativity, in answer to the fundamental questions posed for all artists of the last century. In Xenakis's words, this represents:

the effort to make "art" while "geometrizing", that is, by giving it a reasoned support less perishable than the impulse of the moment, and hence more serious, more worthy of the fierce fight which the human intelligence wages in all the other domains.

Part IV

The composer speaks

ORGANUM

CHAPTER 9

Microtones and projective planes

Carlton Gamer and Robin Wilson

Although most music is composed in the 12-tone equal-tempered system,
attention has also been paid to the systems obtained by dividing the octave
into other numbers of divisions. For certain of these systems, there is an
unexpected connection between the compositional operation of musical
inversion and the idea of 'duality' for certain geometrical objects called finite
projective planes.

For hundreds of years mathematicians and musicians have been
intrigued by the musical systems obtained when an octave is divided up,
not into the usual twelve tones with which we are all familiar, but into
a smaller or larger number of tones. Certain of these systems, such as
the 19-tone, 31-tone and 53-tone equal-tempered systems, have been
much investigated, since they give rise to tunings that more closely
approximate particular intervals in the harmonic series ('just' tunings)
than does the 12-tone equally tempered system; a table comparing
these tunings is given below.

interval	*just ratios*	*12-tone*	*19-tone*	*31-tone*	*53-tone*
octave	2.000	2.000	2.000	2.000	2.000
perfect fifth	1.500	1.498	1.494	1.496	1.500
perfect fourth	1.333	1.335	1.339	1.337	1.333
major third	1.250	1.260	1.245	1.251	1.249
minor third	1.200	1.189	1.200	1.196	1.201

The 19-tone and 31-tone equally tempered systems date from the sixteenth century and were studied by such mathematicians as Marin
Mersenne, who designed a 31-tone keyboard (see Chapter 1), and
Christiaan Huygens, who used logarithms to perform the necessary
numerical calculations. The 53-tone system was studied by Boethius,
Mersenne and others, and a version of it was confirmed as the official
musical system in China in 1713, although a method of equal temperament had already been introduced there by Prince Chu Tsai-yü in 1584,
fifty years before the first writings on the subject in Europe. Indeed, it

An example of 31-tone music: the beginning
of Carlton Gamer's *Organum*, from *Canto
LXXXI* (Ezra Pound).

has been claimed that the idea of equal temperament was familiar to the Chinese by the year 1000.

Systems with fewer than twelve tones are also of interest, although most of these are not equally tempered. For example, 7-tone systems and other systems of a similar nature have been used in the music of India and Thailand, and by the gamelan orchestras of Indonesia. Similarly, the various modes (Dorian, Phrygian, etc.) were much employed in medieval and Renaissance music, from which our own major scale derives.

In the past few years composers have increasingly become involved with 'microtonal' systems. This interest may be due, at least partly, to the desire in the various arts to 'return to fundamentals', as exemplified in the early twentieth century by the serialism of Schoenberg's 'twelve-tone row' (see Chapter 8), the paintings of Mondrian, the sculptures of Brancusi, and the contributions to the foundations of mathematics by Russell and Whitehead. The artistic desire to return to fundamentals has been coupled with a search for new technical and expressive resources, and significant aspects of such resources—the various discoveries feeding into so-called 'geometrical abstract art', the science of digital imaging, the shaping of metals in recent architecture—have been informed in one way or another by mathematics. In consequence, the study of atonal and microtonal music has become increasingly mathematical, involving set theory, permutation groups and, recently, cyclic designs.

Such studies have been given impetus by the advent of electronically produced music. This has made it possible to perform music in systems other than the 12-tone system with far greater accuracy of intonation than previously. In view of this, it seems possible that attention will move away from 'traditional' microtonal systems, such as the 19-tone and 53-tone systems, and that other systems will figure more prominently in the future. In this chapter our interest lies in equal-tempered systems with n tones, where n is a number of the form $k^2 - k + 1$, for some integer k; as we shall see, these numbers arise naturally out of geometrical considerations. Included in this list are the 7-tone and 31-tone systems (corresponding to $k = 3$ and $k = 6$), as well as the less familiar 13-tone and 21-tone systems (corresponding to $k = 4$ and $k = 5$); in all these systems we employ cyclic designs and finite projective planes, concepts that we introduce later. Equally tempered systems with 19 and 53 tones do not fit directly into this classification.

Equally tempered systems

Consider the piano keyboard opposite, depicting the twelve notes of the octave. In order that music in any key can be played as nearly in tune as possible, the tuning is *equally tempered*, so that (for example) $D^\sharp = E^\flat$, $B^\sharp = C$, and $B^\flat = A$. As we saw in Chapter 1, this tuning is effected by

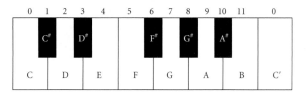

making the ratio of the frequencies of any two consecutive notes equal to $2^{1/12}$ ($= 1.05946\ldots$); note that an octave has frequency ratio 2.

In what follows we are concerned exclusively with *pitch classes*, rather than with actual pitches; this means that all notes with the same letter name (for example, all $G^\sharp$s) are to be regarded as the same. Under this assumption each note can be regarded as an integer modulo 12, and the intervals between notes are obtained by subtraction modulo 12; modular arithmetic is explained in the box.

Modular arithmetic

Two numbers a and b are **congruent modulo n**, written $a \equiv b$ (modulo n), if $a - b$ is divisible by n; for example, $17 \equiv 5$ (modulo 12). The different integers modulo n are usually taken to be $0, 1, 2, \ldots, n - 1$.

Addition or subtraction of integers modulo n is effected by adding or subtracting in the usual way and then determining the remainder on division by n; for example, $8 + 9 \equiv 5$ (modulo 12) and $5 - 9 \equiv 8$ (modulo 12).

Arbitrarily choosing $C = 0$, we obtain the following correspondence:

C	C$^\sharp$	D	D$^\sharp$	E	F	F$^\sharp$	G	G$^\sharp$	A	A$^\sharp$	B
0	1	2	3	4	5	6	7	8	9	10	11

We refer to these notes (numbers) as *tones*, and to this equally tempered system as *ETS 12*.

More generally, we define the *equally tempered system ETS n* to be the system that arises when the octave is divided into n tones in such a way that the ratio of any two successive frequencies is $2^{1/n}$. In such a system the n tones correspond to the integers $0, 1, \ldots, n - 1$ (modulo n), and the intervals between tones are obtained by subtraction modulo n.

In all these systems we can define the musical operation of transposition. Given any set S of tones in *ETS n* we may apply the *transposition* T_k, defined by

$$T_k(x) \equiv x + k \text{ (modulo } n), \quad \text{for all tones } x \text{ in } S;$$

we can think of transposition as playing the same tune but starting on a different note. For example, if $n = 12$ and S is the C major scale, we have

C major:	C	D	E	F	G	A	B	C
S:	0	2	4	5	7	9	11	0

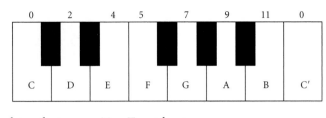

Applying the transposition T_4, we have

E major:	E	F♯	G♯	A	B	C♯	D♯	E
$T_4(S)$:	4	6	8	9	11	1	3	4

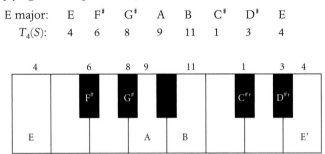

(Beethoven used this particular transposition for the second theme in the first movement of his *Waldstein Sonata*, Op. 53.)

We will be concerned with the number of tones in common between such scales. In this case, note that S and $T_4(S)$ have exactly three tones (4, 9 and 11) in common. These three tones arise from the existence of three intervals of size 4 (0–4, 5–9 and 7–11) in the original scale. More generally, there is an elementary but useful result, first formulated by the composer Milton Babbitt and now known to music theorists as *Babbitt's theorem*:

Given any set of tones, the multiplicity of occurrence of a given interval in the set determines the number of tones in common between that set and its transpositions by that interval.

Expressed symbolically, this says that if S is any set of tones in *ETS n*, then the number of elements common to S and $T_k(S)$ is equal to the number of pairs a, b in S for which $a - b \equiv k$ (modulo n).

In a similar way, we can investigate the musical operation of inversion. Given any set S of tones in *ETS n* we may apply the *inversion I*, defined by

$$I(x) \equiv n - x \text{ (modulo } n), \text{ for all tones } x \text{ in } S;$$

we can think of inversion as reflecting the tones in S vertically in the line C ($= 0$). For example, if $n = 12$, then the inversion of $S = \{2, 7, 11\}$ is

$$\{12 - 2, 12 - 7, 12 - 11\} = \{10, 5, 1\},$$

as shown below.

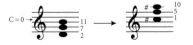

Difference sets

Consider the following extract from the last movement of Béla Bartók's
Fourth String Quartet:

The set of tones involved in this extract is $\{1, 3, 6, 7\}$. This four-tone
set has been used by several composers (for example, in Schoenberg's
Opus 33a, Webern's *Opus 5* and Elliott Carter's *First String Quartet* and
Double Concerto), and is called an *all-interval tetrad* since every possible
interval occurs in it; for example, intervals of size 4, 6 and 10 occur
between the pairs 3 and 7, 1 and 7, and 3 and 1, respectively. In fact, this
tetrad gives rise to each interval exactly once, with the single exception
of the tritone interval of size 6 which can be written in two ways, as
$1 - 7$ or $7 - 1$ (modulo 12).

It would be even more satisfactory if every possible interval, *without
exception*, were to occur just once. This leads us to investigate sets of
tones in an equally tempered system that have this property. With this in
mind, we introduce the idea of a *difference set (modulo n)* to be a set of
distinct integers $c_1, \ldots, c_k$ (modulo n) for which the differences $c_i - c_j$
(for $i \neq j$) include each non-zero integer (modulo n) exactly once; for
example:

- $\{0, 1, 3\}$ is a difference set (modulo 7), since the differences are
  $1 \equiv 1 - 0, 2 \equiv 3 - 1, 3 \equiv 3 - 0, 4 \equiv 0 - 3, 5 \equiv 1 - 3$ and $6 \equiv 0 - 1$;
- $\{0, 4, 6\}$ is also a difference set (modulo 7);
- $\{1, 3, 6, 7\}$ is a difference set (modulo 13), but (as shown above) it is
  not a difference set (modulo 12) because the 'tritone difference'
  occurs twice;
- $\{0, 1, 4, 6, 13, 21\}$ and $\{0, 10, 18, 25, 27, 30\}$ are both difference sets
  (modulo 31).

Cyclic designs

We shall also need the concept of a cyclic design. Given positive integers
n and k, with $k < n$, a *cyclic design* with these parameters is an arrangement of n numbers into n blocks of size k, in such a way that any two
numbers appear together in exactly one block, and that the numbers in
each successive block are obtained from those of the previous one by
adding 1 (modulo n). For example, a cyclic design with parameters
$n = 13$ and $k = 4$ is as follows, with the blocks written vertically; you can
check that any pair of numbers (such as 6 and 10) appear together in exactly

one block—in this case, block (4)—and that the numbers in each successive block are obtained from those of the previous one by adding 1 (modulo 12).

(1)	(2)	(3)	(4)	(5)	(6)	(7)	(8)	(9)	(10)	(11)	(12)	(13)
1	2	3	4	5	6	7	8	9	10	11	12	0
3	4	5	6	7	8	9	10	11	12	0	1	2
6	7	8	9	10	11	12	0	1	2	3	4	5
7	8	9	10	11	12	0	1	2	3	4	5	6

In musical terms, the numbers 0, 1,..., 12 appearing in this cyclic design can be thought of as the tones in the equally tempered system *ETS 13*, and the successive blocks can be thought of as tetrads that are obtained from earlier ones by transposition. Note that any two tones in *ETS 13* appear together in just one tetrad—for example, the tones 6 and 10 appear together in tetrad (4). Note also that 4×3, the number of possible differences between numbers in the difference set, is equal to 12, the number of non-zero integers modulo 12.

More generally, we can construct such cyclic designs whenever we have a difference set. For, if S is a difference set (modulo n), then S gives rise to a cyclic design whose first block is S and whose successive blocks are obtained by adding 1 (modulo n) to each element of the preceding block. For example, the difference set $\{1, 3, 6, 7\}$ (modulo 13) gives rise to the cyclic design above.

It follows from the definition of a difference set (modulo n) that $k(k-1)$, the number of possible differences between two numbers in the difference set, must be equal to $n-1$, the number of non-zero integers modulo n. Thus, difference sets (modulo n) can occur only when $k(k-1) = n-1$, for some integer k—that is, $n = k^2 - k + 1$.

Finite projective planes

A *finite projective plane* is a geometrical system consisting of a finite number of points and lines, with the properties that any two points lie on just one line, and any two lines pass through just one point. In such a system it can be shown that each line contains exactly k points and that each point lies on exactly k lines, for some integer k, and that in total there must be exactly $k^2 - k + 1$ points and $k^2 - k + 1$ lines. For example, the following finite projective plane corresponding to $k = 3$ has exactly 3 points lying on each line, and exactly 3 lines passing through each point. Since $3^2 - 3 + 1 = 7$, there are exactly 7 points (0, 1, 2, 3, 4, 5, 6) and 7 lines ((0), (1), (2), (3), (4), (5), (6)) in this system; the reason for labelling the lines in this way will become apparent soon. It is often called the *Fano plane*, since it was introduced by the Italian geometer Gino Fano, in 1890; notice that one of the lines has to be drawn curved, but this does not invalidate the concept.

More complicated is the finite projective plane corresponding to $k = 4$; since $4^2 - 4 + 1 = 13$, this has 13 points and 13 lines, with 4 points lying on each line and 4 lines passing through each point.

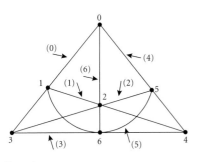

Fano plane.

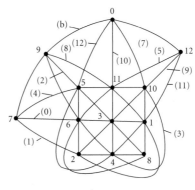

13-point projective plane.

We have seen that any difference set in *ETS n* leads to a cyclic design with $n = k^2 - k + 1$ tones, in which any two tones appear together just once. For many values of k, such a cyclic design gives rise to a finite projective plane. We now look at a difference set in each of the systems *ETS 7*, *ETS 13* and *ETS 31*, and obtain the corresponding cyclic designs and finite projective planes. However, although $43 = 7^2 - 7 + 1$, the equally-tempered system *ETS 43* cannot be studied in this way, since it can be proved that there exists no projective plane with $n = 43$.

ETS 7 *Fano plane*: difference set = the triad {0, 1, 3}

(0)	(1)	(2)	(3)	(4)	(5)	(6)
0	1	2	3	4	5	6
1	2	3	4	5	6	0
3	4	5	6	0	1	2

ETS 13 *13-point projective plane*: difference set = the tetrad {1, 3, 6, 7}

(0)	(1)	(2)	(3)	(4)	(5)	(6)	(7)	(8)	(9)	(10)	(11)	(12)
1	2	3	4	5	6	7	8	9	10	11	12	0
3	4	5	6	7	8	9	10	11	12	0	1	2
6	7	8	9	10	11	12	0	1	2	3	4	5
7	8	9	10	11	12	0	1	2	3	4	5	6

ETS 31 *31-point projective plane*: difference set = the hexad {0, 1, 4, 6, 13, 21}

(0)	(1)	(2)	(3)	(4)	(5)	(6)	(7)	(8)	(9)	(10)	...	(29)	(30)
0	1	2	3	4	5	6	7	8	9	10	...	29	30
1	2	3	4	5	6	7	8	9	10	11	...	30	0
4	5	6	7	8	9	10	11	12	13	14	...	2	3
6	7	8	9	10	11	12	13	14	15	16	...	4	5
13	14	15	16	17	18	19	20	21	22	23	...	11	12
21	22	23	24	25	26	27	28	29	30	0	...	19	20

Duality

One of the most important ideas in geometry is that of *duality*, in which we can obtain a new system by interchanging the roles of points and lines. In particular, for every finite projective plane there is a *dual plane* obtained by interchanging the points and lines; thus, if *p* is a point and *l* is a line of the original projective plane, then *l* is a point and *p* is a line of the dual plane. *Lines through the point p* of the original plane then become *points lying on the line p* in the dual plane, and *points lying on the line l* in the original plane become *lines passing through the point l* in the dual plane. For example, the dual plane of the Fano plane is as follows; note that the lines (0), (4), (6) passing through the point 0 in the original plane become the points (0), (4), (6) lying on the line 0 in the dual plane, and similarly for the other points and lines.

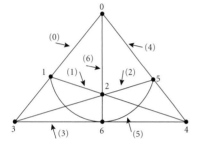

Fano plane.

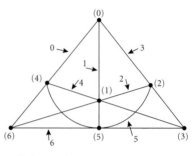

Dual of Fano plane.

We now return to our finite projective planes for the systems *ETS 7*, *ETS 13* and *ETS 31*, and find their dual planes. For the system *ETS 7*, the dual plane also turns out to be a cyclic design, and the numbers 0, 4, 6 appearing in its first block {(0), (4), (6)} can also be obtained by subtracting from 7 the numbers in the original difference set {0, 1, 3}. For the system *ETS 13*, the dual plane is again a cyclic design, and the numbers 6, 7, 10 and 12 appearing in its first block {(6), (7), (10), (12)} can be obtained by subtracting from 13 the numbers in the original difference set {1, 3, 6, 7}. Similarly, for the system *ETS 31*, the dual plane is a cyclic design and its first block is obtained by subtracting the numbers in the original difference set from the number *n* of points. The dual plane for *ETS 31* is of musical interest, since the first four notes of each hexachord (for example, (0), (10), (18), (25)) form a perfect dominant 7th chord, as shown on the following 31-tone keyboard.

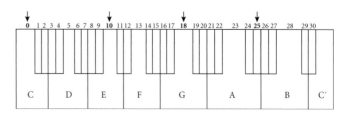

ETS 7 Fano plane:

(0)	(1)	(2)	(3)	(4)	(5)	(6)
0	1	2	3	4	5	6
1	2	3	4	5	6	0
3	4	5	6	0	1	2

Dual plane:

0	1	2	3	4	5	6
(0)	(1)	(2)	(3)	(4)	(5)	(6)
(4)	(5)	(6)	(0)	(1)	(2)	(3)
(6)	(0)	(1)	(2)	(3)	(4)	(5)

ETS 13 Finite projective plane:

(0)	(1)	(2)	(3)	(4)	(5)	(6)	(7)	(8)	(9)	(10)	(11)	(12)
1	2	3	4	5	6	7	8	9	10	11	12	0
3	4	5	6	7	8	9	10	11	12	0	1	2
6	7	8	9	10	11	12	0	1	2	3	4	5
7	8	9	10	11	12	0	1	2	3	4	5	6

Dual plane:

0	1	2	3	4	5	6	7	8	9	10	11	12
(6)	(7)	(8)	(9)	(10)	(11)	(12)	(0)	(1)	(2)	(3)	(4)	(5)
(7)	(8)	(9)	(10)	(11)	(12)	(0)	(1)	(2)	(3)	(4)	(5)	(6)
(10)	(11)	(12)	(0)	(1)	(2)	(3)	(4)	(5)	(6)	(7)	(8)	(9)
(12)	(0)	(1)	(2)	(3)	(4)	(5)	(6)	(7)	(8)	(9)	(10)	(11)

ETS 31 Finite projective plane:

(0)	(1)	(2)	(3)	(4)	(5)	(6)	(7)	(8)	(9)	(10)	...	(29)	(30)
0	1	2	3	4	5	6	7	8	9	10	...	29	30
1	2	3	4	5	6	7	8	9	10	11	...	30	0
4	5	6	7	8	9	10	11	12	13	14	...	2	3
6	7	8	9	10	11	12	13	14	15	16	...	4	5
13	14	15	16	17	18	19	20	21	22	23	...	11	12
21	22	23	24	25	26	27	28	29	30	0	...	19	20

Dual plane:

0	1	2	3	4	5	6	7	8	9	10	...	29	30
(0)	(1)	(2)	(3)	(4)	(5)	(6)	(7)	(8)	(9)	(10)	...	(29)	(30)
(10)	(11)	(12)	(13)	(14)	(15)	(16)	(17)	(18)	(19)	(20)	...	(8)	(9)
(18)	(19)	(20)	(21)	(22)	(23)	(24)	(25)	(26)	(27)	(28)	...	(16)	(17)
(25)	(26)	(27)	(28)	(29)	(30)	(0)	(1)	(2)	(3)	(4)	...	(23)	(24)
(27)	(28)	(29)	(30)	(0)	(1)	(2)	(3)	(4)	(5)	(6)	...	(25)	(26)
(30)	(0)	(1)	(2)	(3)	(4)	(5)	(6)	(7)	(8)	(9)	...	(28)	(29)

Inversion

In each of these examples the blocks of the dual plane can be obtained by subtracting from n (the number of tones) the numbers in the original difference set: this construction corresponds to musical inversion. Up to now, this direct link between the concepts of geometrical duality and musical inversion—that *the dual plane corresponds precisely to inversion of the tones in the original difference set*—has been largely unnoticed by musicians; for a proof of this result, see the following box. It is our hope that musicians will find it worthwhile to concentrate increasingly on equally tempered systems *ETS* n, where $n = k^2 - k + 1$ for some integer k, and that they will find the concept of duality to be fruitful in both musical analysis and composition.

Theorem. *The dual plane corresponds precisely to inversion of the original difference set*

Proof. Let the original difference set (block (0)) be $\{c_1, \ldots, c_k\}$. We must show that the dual plane is a cyclic design whose first block is $n - c_1, \ldots, n - c_k$.

Note first that $\{n - c_1, \ldots, n - c_k\}$ is a difference set, since if $s = c_i - c_j$, then $s = (n - c_j) - (n - c_i)$, and so there is a one-one correspondence between the differences formed by the two sets. Note also that tone 0 occurs in blocks $n - c_1, \ldots, n - c_k$, and so these numbers form the first block of the dual plane. Similarly, each other tone t occurs in blocks $n - c_1 + t, \ldots, n - c_k + t$, and so these numbers form a block of the dual plane. It follows that the dual plane is a cyclic design whose first block is $n - c_1, \ldots, n - c_k$.

Fanovar: Variations on a Fano plane

We conclude by describing the composition *Fanovar* by the first author.

As its title indicates, the piece is governed by the structure of the Fano plane. It is composed for seven instruments and consists of seven sections, or variations, of which the first two are presented here. The seven instruments are grouped into seven trios in accordance with the disposition of points and lines in the Fano plane, as follows.

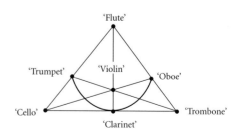

Each of the seven instruments is to be tuned to play in the equally tempered system *ETS 7*, and for this reason, the most feasible realization of the score entails the use of electronically synthesized instruments. The quotation marks around the name of each instrument in the score are intended to suggest a traditional instrumental timbre that approximates that of the instrument shown. Furthermore, because of the temperament employed, the lines and spaces in the staves of the score do not denote the scale degrees of the traditional 'white-key' subcollection of *ETS 12*; rather, they denote the scale degrees of *ETS 7* with pitch classes C = 0, D = 1, E = 2, F = 3, G = 4, A = 5 and B = 6.

The melodic content of the piece is governed by a different seven-note diagram, in which each point represents a pitch name, such as C. The diagram for the first variation (see page 160) is shown below. Note that the instrumental triad of 'cello'–'trombone'–'clarinet' plays the pitch classes C, D, G that appear in the triangle on the left, while the 'oboe' and 'trumpet' play the pitch classes A, B, F, E that form a path on the right.

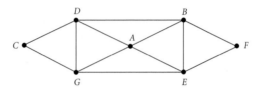

The second variation (see page 161) follows similar instrumentational and melodic principles, with the diagram re-lettered as shown below; for example, the instrumental triad of 'flute'–'clarinet'–'violin' plays the pitch classes G, B, D that appear in the triangle on the left.

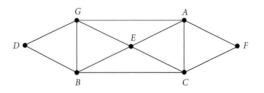

Fanovar

Carlton Gamer
(1996)

First variation of *Fanovar.*

B Variation 2

A little slower

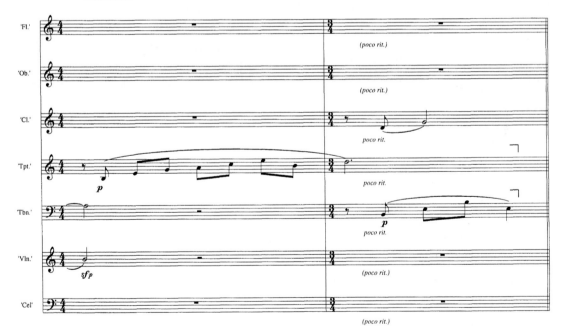

Second variation of *Fanovar*

CHAPTER 10

Composing with fractals

Robert Sherlaw Johnson

An iterative formula can be used to generate two-dimensional patterns on a plane. A computer program is described which generates musical patterns using the same principles leading to a completed composition.

A glance through the illustrations in *The science of fractal images* edited by Peitgen and Saupe, or *The beauty of fractals* by Peitgen and Richter is sufficient to suggest that it is possible to create unusual and interesting patterns, and even landscape and floral pictures, by computer generation. These works are concerned principally with the visual application of fractals, music receiving only a brief mention in the former book. One is tempted to speculate, however, that if meaningful visual patterns can be created by fractal generation, then it should also be possible to create aural ones. This chapter is the result of such an investigation using one particular type of fractal generation.

Of the various fractal sets described in the above literature and elsewhere, the one that seems to have caught the popular imagination is the Mandelbrot set, shown on the left. It has been the subject of a number of computer programs, designed to create the visual image of this set on the screen, largely because of the large variety of detail that appears when one zooms in on specific parts of the 'picture'. Some use has been made by others of the Mandelbrot set to generate music, but none of the results that have come to the knowledge of the author has seemed particularly impressive. Others have used the concepts of composition derived from chaos theory and fractals—for example, the self-similarity of nesting one musical phrase within itself, each note of the sequence generating the sequence itself at different pitches.

The method with which we are concerned here, however, is not the Mandelbrot set, but a chaotic dynamical system such as described in Peitgen and Saupe's book. The actual iterative formula employed originated with Martin Bell of Aston University, and appeared in A. K. Dewdney's mathematics column in *Scientific American* in 1986. It was designed to create interesting and symmetrical patterns on the two-dimensional computer screen coordinatized by two variables

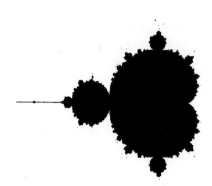

The Mandelbrot set.

Robert Sherlaw Johnson.

x and *y*:

$$x \leftarrow y - \mathrm{sign}\,(x)\,\sqrt{|bx - c|}, \quad y \leftarrow a - x.$$

In the first of these formulas, sign (x) means that 1 is substituted for this if x is positive, -1 if x is negative, and 0 if x is zero; $a,\ b$ and c are any real number constants. The initial values for the variables x and y are usually 0, but can be chosen to be any real numbers. The formula on the right-hand side of each arrow is calculated in turn and assigned to the variable on the left: these new values are then substituted into the right-hand side and new values for x and y are calculated, and so on.

The result is a string of real numbers which are then converted to points on the computer screen, so building up a pattern over a period of time. Using this process as the basis, a computer program was devised (written in Modula-2) on the Atari ST computer to convert the variables to sound, while still retaining the plot facility.

Unlike the flat surface of the screen, music is a multi-dimensional environment, these dimensions consisting of pitch, duration, the time interval between successive sounds, timbre, loudness, tempo, and so on. The most obvious assignment of the x- and y-values would be pitch and the time interval between successive sounds (rhythm), but although early experiments proved to be promising as regards pitch, they were not so as regards rhythm, because of the different way in which rhythmic patterns are built up. A melodic pattern can consist of a large number of different pitches but a smaller number of different durations, which have to form some kind of metrical pattern (although this may be quite complex), in order to make rhythmic sense. In addition, rhythm is not only a question of durational patterns, but also of accentual ones. Initially it was decided, therefore, to work on constant streams of pitches and to assign the y-values to loudness. As it happened, a sense of rhythm emerged through the interaction of accented (louder) sounds with pitch. The pitch information is distributed among eight channels of a synthesizer (as described below), so that selection of one or two channels can also create rhythmic patterns of different durations.

Translating real numbers into sound

Real numbers could be translated into pitch in a variety of ways, but the most convenient, using a computer with a MIDI output port, was to drive a synthesizer capable of receiving MIDI information. As MIDI devices recognize integers and not real numbers, the x- and y-values had to be scaled to be within the acceptable range for pitches and loudness (0 to 127). It was not desirable that the whole pitch-range should be used all the time, so the program was designed to allow the user to decide the range to be used in any given instance. Initially, then, the user chooses real numbers that represent an index for range of pitch and one for range of dynamics, and an integer representing tempo. All these values can be changed in real time while the program is running. The real indices representing pitch and dynamic ranges start from 0 (representing no variation) and vary upwards in steps of 0.1. The precise effect of any given index on pitch would depend on the spread of x-values generated, and would need to be adjusted by trial and error in any given case for the most satisfactory results.

One of the major problems was how to drive the eight MIDI channels available on the synthesizer, and how to assign 'voices' (timbres) to them. The latter problem was accomplished by trial and error, on the basis of providing a mixture of percussive and sustained sounds in order to achieve contrast. The selection of channel for any given pitch generated was built into the computer program.

At this point it should be noted that all these parameters: pitch and dynamic range, tempo and channel selection, could have been brought under fractal control by extending the formula to include more than two variables, or by further calculations from the formula involving more constants. For musical reasons, rapid changes of these parameters is not desirable, so that (for the present) they were left under user control. Selection of channel was tried using the y-values of the formula, as well as by a three-dimensional extension, but the rapid change of timbres involved tended to produce a monotony because all eight channels were equally favoured continually. The method eventually implemented was to assign ranges of values of y (rounded to four decimal places) to each channel. Although these ranges can also be decided by the user, the program provides default settings, as shown in the following table:

channel:	1	2	3	4	5	6	7	8
values of $\|y\|$:	0–1.9999	2–3.9999	4–5.9999	6–6.9999	7–7.9999	8–8.9999	9–9.9999	10 or over

If low values are selected for the constants, the corresponding values of y tend to be in the lower ranges, which means that values above 10 are not particularly frequent; this results in an overall balance between the eight channels.

In some cases, there is only a gradual expansion of values of y around 0, which gives rise to a sense of 'orchestration', in so far as the higher numbered channels are introduced progressively. In other cases (including the sequence for *Fractal in A flat*), there are subsequent contractions in the selection of channels, as well as an initial expansion, providing even greater variety in the range of timbres. All these ranges can be altered at the start to accommodate different spreads of the y-value. This can even be taken to the extreme case of excluding certain channels by giving them a range of 0, or limiting the number of channels by setting abnormally high ranges for the upper ones. For example, the program asks for upper limits of the range of y-values for each channel: settings of 0, 2, 4, 4, 6, 8, 1000 will silence channels 1 (range 0 to 0), 4 (range 4 to 4) and 8, except in the unlikely event of a value of y appearing greater than 1000; 1000 being the upper limit of channel 7 is automatically the lower limit of channel 8.

So far, each pitch is sustained until silenced by the next one, so that some means of allowing simultaneity of sound had to be devised. This was achieved by choosing another constant representing the maximum duration that a sound can have. This constant is divided into the note's MIDI-integer and the remainder assigned to a counter which is decremented on each subsequent generation of x and y: the note is stopped when the counter reaches 0. There is one modifying factor that results from the way in which the synthesizer handles MIDI information: when another note is assigned to the same channel, the previous note is cut short, even if 0 has not been reached on the counter. This means that, although initially each note will have the same potential duration, in some cases it may not reach its full length, owing to the arrival of a new note on the same channel.

One respect in which a musical interpretation of the formula differs radically from a visual one is the ability to create different patterns by selecting only alternate values of x and y, or every third value, and so on. As the visual pattern is built up on the screen, one sees the accumulation of points created by all the values generated up to that particular moment. The longer the generation takes place, the more interesting and complex the pattern becomes. Music is not perceived in this way. One hears patterns that occur at the moment of listening, and one may perceive that these relate to something that happened earlier, or that they are different. For visual realization it is only the whole accumulation of values that gives rise to a sensible pattern, whereas in the case of music the whole is not perceived simultaneously and only localized patterns make sense. It is easy to perceive, therefore, why rejecting particular generations from the stream can create radically different musical patterns, whereas it makes no difference to the cumulative effect of the visual pattern, except to make it sparser in appearance. *Example 1* opposite illustrates this: in (a) all the values are converted to pitches, whereas in (b) every third value is converted; although only fragments, the difference of melodic behaviour can easily be seen.

It is a characteristic of this particular iterative formula that, for most low-value constants, values of x and y remain within a reasonable range for at least the first few thousand generations. The screen-patterns generated show a clear ordering because the dots tend to cluster in certain areas, leaving other areas blank. In an extreme case ($a = 8$, $b = 4$, $c = 0$), two interweaving lines are generated, as shown on the left. These lines are not caused by a single row of dots, but by narrow clusters of dots,

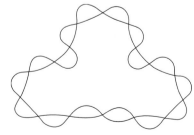

two interweaving lines
($a = 8$, $b = 4$, $c = 0$)

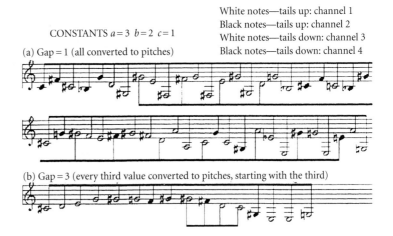

CONSTANTS $a = 3$ $b = 2$ $c = 1$

White notes—tails up: channel 1
Black notes—tails up: channel 2
White notes—tails down: channel 3
Black notes—tails down: channel 4

(a) Gap = 1 (all converted to pitches)

(b) Gap = 3 (every third value converted to pitches, starting with the third)

Example 1.

so that an exact placing of a particular dot is not predictable: all that can be said is that they are attracted towards a line, which I called an *attractor*.

In most cases, dots tend to be attracted to particular areas, rather than form in narrow lines, and it is largely this feature that allows such sequences to be recognised as patterns, whether visual or musical.

Composing with the formula

A variety of patterns can be generated involving recognizable repetitions and transformations, but the raw sequences from any particular constants are too uninteresting in the long run without introducing the human element in the form of the composer. The question of how much a composer should interfere with the process of generation is not an easy one to answer, as in an extreme case it would be possible to cut and paste different sequences to the extent that the fractal generation element becomes degraded. Imagine a hypothetical case where a composition is assembled from fragments of sequences involving different sets of constants, each fragment involving perhaps no more than about 30 generations of x and y. It could be argued in such a case that the fractal element has been broken up to the extent that the resulting composition could not be fairly called 'fractal'. It is not easy to see what the allowable extent of interface from the human composer should be, but the composition now to be described perhaps provides some pointers.

Fractal in A flat

During the process of experimenting with different constants and other parameters, one particular combination displayed markedly tonal characteristics as well as producing other usable motivic ideas. The three constants used for the formula were $a = 1$, $b = 0.1$ and $c = 1$, and the index for pitch spread was 2. This latter number is critical, in that any variation destroys the sense of A flat tonality that arises from the predominance of the major triad in the opening sequence, as illustrated below in *Example 2*.

Tails up: channel 1; tails down: channel 2

Example 2.   * Channel 2 moves from the C to the second E flat after the eighth time.

The index for the dynamic range is less critical, provided that it is not too large to silence the notes at the lower end of the dynamic scale. The integer representing tempo has to vary with the computer running the program. The ST on which the program was initially developed, and on which this composition was generated, took a tempo index of 110 (the larger the index, the slower the tempo). On the Atari TT, however, an index of around 600 is needed to generate an approximately similar tempo. Other crucial parameters are an index of 16 for the maximum duration of pitches, and 2 for the gap value. The choice of 16 ensures a reasonable variation of long and short notes, and the gap value of 2 means that every other value of x (starting with the first) is converted to pitch.

An interesting characteristic of the basic sequence is the way in which a distinct musical shape develops. For the channel selection, the above default settings were used. After a prolonged 'duet' involving the first two channels, the remaining channels quickly enter, giving a sense of development in the music. After a while, the 'duet' returns—more prolonged this time—followed by the rapid re-entry of the other channels. This creates a binary structure A–B–A–B in the music—the following diagram shows the shape.

Channel 3 on its own also presented some interesting features, as shown below in *Example 3*. It is not involved in the initial 'duet' and starts on generation 828. A natural musical development was evident from this channel, involving repetition, transformation and contrast. Perhaps the most surprising feature was the return of the opening motif (a) at the end of the sequence, creating once again a sense of musical form.

Example 3.

The composition *Fractal in A flat* was constructed in the following way. Five minutes of the sequence derived from the constants $a = 1$, $b = 0.1$ and $c = 1$ were taken as the 'base' sequence in *Example 2*. Channel 3 alone was directed to a flute voice on the synthesizer (*Example 3*), and part of it was superimposed on the first 'duet' of the base sequence, and the whole of it at a later stage. Other sequences were also extracted in the same way for superimposition, as follows.

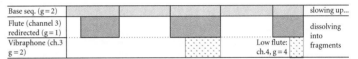

Base seq. (g = 2)						slowing up...
Flute (channel 3) redirected (g = 1)						dissolving into
Vibraphone (ch.3 g = 2)					Low flute: ch.4, g = 4	fragments

The voice names are those for the appropriate channels on the synthesizer; g = gap.

The more interesting sections of the base sequence were left to stand on their own, except for the final section where a more complex and cumulative effect was sought towards the end of the piece. The problem of ending the piece (as, in theory, a sequence is infinite there was no natural ending to the base sequence) was achieved by increasing the gap so that the slowing up of tempo as well as a change in musical character were achieved. At the same time, fragments of other tracks were superimposed, finishing with an open flute bar from *Example 3*.

The 'butterfly' effect

It is difficult to predict what sort of visual pattern will arise from a given set of constants. They tend to divide themselves, however, into distinct groups with similar characteristics, three of which are illustrated overleaf. Each of these patterns has been captured from the screen-plot generated by the fractal program after half a minute, (about 15 000 generations) without music conversion (so that it runs at its maximum speed) on an Atari TT.

It is possible also to associate particular musical characteristics with particular sets of constants if the gap is set to 1, but it is more difficult to identify these by name because of the inherently abstract nature of the music.

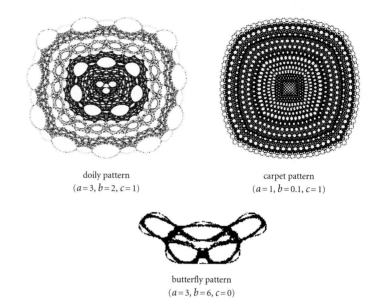

doily pattern
$(a=3, b=2, c=1)$

carpet pattern
$(a=1, b=0.1, c=1)$

butterfly pattern
$(a=3, b=6, c=0)$

It is said that if a butterfly flaps its wings in Venezuela, this can cause a hurricane in Florida. The quotation may not be entirely accurate and it may be impossible to verify, but it emphasizes the unpredictability of dynamical systems that govern such things as the weather or the economy. A feature of all these systems is that a very slight variation in one of the constants can produce quite markedly different results in the variables as they develop. In the case of the constants used for *Fractal in A flat*, slight variations in the constants produce patterns of the same type, but with noticeably different features, as illustrated below.

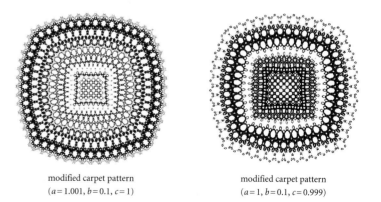

modified carpet pattern
$(a=1.001, b=0.1, c=1)$

modified carpet pattern
$(a=1, b=0.1, c=0.999)$

Using these sets of constants, the music does not initially vary until generation 656 of x and y, when they begin to diverge from the original version, and at generation 1566 when they diverge from each other. In spite of these divergences, they return to similar patterns spasmodically at a later stage. In actual fact, divergences in the values of x take place almost immediately, but they are too small to affect the conversion of these real values into 'note' integers, and they are also too small for there to be an immediate difference in the visual pattern.

Fractal dialogue (Variations)

In order to illustrate the versatility of the programme, a second composition, *Fractal dialogue (Variations)*, was designed to exploit melodic variations obtainable from varying the channel and pitch ranges. The generating constants were $a = 0.05$, $b = 0.5$ and $c = 5$. The visual pattern generated is less interesting than the one that corresponds to *Fractal in A flat*, but one of the notable features of this formula is the general lack of correlation between the visual and audible aspects as far as interest is concerned. The principal line of music alternates between oboe and flute sound and employs only the first 171 generations on channel 2, although this is extended to the first 285 generations for the last two variations. Each variation either moves or expands the y-values used to excite channel 2, while the pitch-range is also varied. This produces a greater elaboration of the initial 'theme' each time, as illustrated in *Example 4*.

Example 4.

What is composed?

In the selection of timbres for the eight channels, there is clearly a compositional choice involved. There is also a compositional choice in selection of sequences and the manner in which they are overlaid, as well as in the manner and point of termination. Also important is the extent to which the computer is allowed to generate the music, rather than the composer—in other words, how long each individual sequence should be. In *Fractal in A flat*, one of these sequences—the solo channel 3 sequence—*was* computer determined, in that there was a natural shape and termination to the sequence as a consequence of the generation. The decision, however, to seek out these features is again a compositional one, rather than arbitrary or dependent in some way on the fractal generation itself. By harnessing several different fractal processes, or by extending the scope of the one described, some (or even all) of these features could be brought under fractal control. This would, of course, need a radical revision of the program. Another course is possible: following the practice of John Cage, decisions about length of sequences, position and number of overlays, etc., could be made the result of chance processes, by the rolling of dice or by random number generation by the computer.

One question remains to be answered: can the computer be said to have 'composed' anything? The whole idea arose from the awareness that fractal sound patterns generated by this means made musical sense. Yet the computer could not be described as 'making decisions' about melodic shape, motivic repetition or form, all of which are apparent in the channel 3 generation in *Fractal in A flat*. If these patterns are apparent to a listener, it can only be because the listener tries to make formal sense of any kind of pattern in sound. (The question of whether this is in itself a creative process is not one that can be gone into here, as it has ramifications outside the scope of the question of fractal music.) This is normally done in response to a composer who has generated these patterns, although not all listeners (depending partly on their musical background) may perceive these patterns as making sense to them. It is possible for naturally occurring sounds to form 'musical' patterns, if only in a rudimentary way. The stream of variables generated by the above formula, however, cannot be included amongst these, as it is only when the stream of variables is harnessed in a particular way that musical or visual sense is derived from it, otherwise it remains a chaotic sequence.

Neil Bibby was formerly Director of 'mathstudio', a consultancy for international mathematics education based in The Netherlands. He held lectureships in mathematics education in UK universities, as well as having taught mathematics in schools in the UK, Italy and the Netherlands. An active researcher in the history of mathematics, he presented many talks related to this research at conferences, especially those of the British Society for the History of Mathematics. He died in 2002.

Jonathan Cross is Lecturer in Music at the University of Oxford and Fellow of Christ Church. He is author of *The Stravinsky legacy* (1998), *Harrison Birtwistle: Man, mind, music* (2000), and editor of *The Cambridge companion to Stravinsky* (2002). He is also editor of the journal *Music analysis*.

John Fauvel was Senior Lecturer in Mathematics at the Open University, and former President of the British Society for the History of Mathematics. He was also involved in an international study of the relationships between the history and pedagogy of mathematics. He was an editor and co-editor of several books, including *Darwin to Einstein: historical studies on science and belief* (1980), *Conceptions of inquiry* (1981), *The history of mathematics: a reader* (1987), *Let Newton be!* (1988), *Möbius and his band* (1993), and *Oxford figures* (2000). He died in 2001.

J. V. Field is Visiting Research Fellow in the School of History of Art, Film and Visual Media at Birkbeck College, University of London. Dr Field's books include *Kepler's geometrical cosmology, The geometrical work of Gerard Desargues* (with J. J. Gray), *Science in art: Works in the National Gallery that illustrate the history of science and technology* (with F. A. J. L. James) and *The invention of infinity: Mathematics and art in the Renaissance*.

Raymond Flood is University Lecturer in Computing Studies and Mathematics at the Department for Continuing Education, Oxford University, and Fellow of Kellogg College. His main research interests lie in statistics and the history of mathematics. He is a co-editor of *The nature of time* (1986), *Let Newton be!* (1988), *Möbius and his band* (1993), and *Oxford figures* (2000).

David Fowler is an Emeritus Reader in Mathematics at the University of Warwick. He has had a long-standing interest in music—its theory, the physics behind it, and what it sounds like, and most particularly, the music for (and performance on) the piano. He is the author of articles on the history of mathematics and other topics, and of the book *The mathematics of Plato's Academy: A new reconstruction* (Oxford, 1987, 1999).

Carlton Gamer is Professor Emeritus of Music at The Colorado College, USA. He is a composer and music theorist, and his compositions have been featured in prominent venues in the United States and Europe. His main theoretical interest is in equal-tempered microtonal systems.

Wilfrid Hodges is Professor of Mathematics at Queen Mary, University of London. His research is in mathematical logic, and in particular model theory and semantics. In his youth he sang as a choirboy at Christ Church Cathedral, Oxford, under Thomas Armstrong, and at the King's School Canterbury he studied the piano with Ronald Smith and the violin with Clarence Myerscough.

Dermot Roaf is Fellow in Mathematics at Exeter College, University of Oxford, and is Captain of the Ringers at St. Giles' Church, Oxford. He has rung most of the methods mentioned in his chapter, including *White's No-Call Doubles*.

Robert Sherlaw Johnson became a lecturer in music at the University of York in 1965. In 1970 he went on to teach at Oxford University, where he continued to work until his death in 2000. His many compositions include *A Northumbrian symphony*, three piano sonatas and an opera: *The Lambton worm*. His book *Messiaen* (1975) remains the definitive work on the composer in English, and he also recorded many of Messiaen's piano works and the song cycles with Noëlle Barker.

Ian Stewart is Professor of Mathematics at the University of Warwick. His research area is bifurcation theory and non-linear dynamics. He is active in the popularization of mathematics, and presented the Royal Institution Christmas lectures in 1997. He is the author of over sixty books, including *The problems of mathematics* (1987) and *Does God play dice?* (1989). He was elected a Fellow of the Royal Society in 2001.

Charles Taylor was Professor of Physics at Cardiff from 1965 to 1983, and Professor of Experimental Physics at the Royal Institution from 1977 to 1989. He presented the Royal Institution Christmas Lectures on television in 1971 (*Sounds of music*) and in 1989 (*Exploring music*) and in 1986 he received the Michael Faraday award for contributions to the Public Understanding of Science. He died in 2001.

Arthur White is professor of mathematics at Western Michigan University, Kalamazoo, Michigan. His researches in topological graph theory have led to several books and articles, and to the composition of three pieces of change-ringing music.

Robin Wilson is Head of the Pure Mathematics Department at the Open University and Fellow of Keble College, Oxford. He has written and edited a number of books on graph theory and combinatorics, including *Introduction to graph theory* (1972, 1996) and *Four colours suffice* (2002). He is involved with the popularization of mathematics and with the history of mathematics, and is a co-editor of *Let Newton be!* (1988), *Möbius and his band* (1993), and *Oxford figures* (2000).

Susan Wollenberg is a Reader in Music at the University of Oxford, as well as Fellow and Tutor of Lady Margaret Hall, and Lecturer in Music at Brasenose College. She has contributed to numerous international conferences and publications. Her monograph on the history of music in Oxford (*Music at Oxford in the eighteenth and nineteenth centuries*) was published by Oxford University Press in 2001.

NOTES, REFERENCES, AND
FURTHER READING

QUOTATIONS

viii   And so they have handed down to us   Archytas of
Tarentum, early fourth century BC.

viii   We must maintain the principle   Plato, *Republic 531*.

viii   The Pythagoreans considered all mathematical
science   Proclus, *Commentary on Book I of Euclid's
Elements*.

viii   This science [mathematics] is the easiest   Roger
Bacon, *Opus Maius* **4**, iii.

viii   I do present you with a man of mine   William
Shakespeare, *The taming of the shrew*, Act 2, Sc. i.

ix   May not Music be described as the Mathematic of
Sense   James Joseph Sylvester, 'Algebraical researches
containing a disquisition on Newton's rule for the
discovery of imaginary roots', *Phil. Trans.* **154** (1865),
613.

ix   Mathematics and music, the most sharply contrasted
fields   H. von Helmholtz, *Vorträge und Reden*
(1884), 82.

ix   Quite suddenly a young violinist appeared   Walter
Heisenberg, *Physics and beyond*, Allen & Unwin (1971), 11.

ix   When Professor Spitta   D. MacHale, *George Boole:
His life and work*, Boole Press, Dublin, 1985.

MUSIC AND MATHEMATICS: AN OVERVIEW

1   *Musicke* I here call that *Science*   John Dee, *The mathe-
maticall praeface to the elements of geometrie of Euclid of
Megara*, 1570.

3   'the most transient Graces' could be 'mathematically
delineated'   See A. Richards, *The free fantasia and the
musical picturesque*, Cambridge (2001), 77–79, for a
discussion of the phenomenon. Burney's description
appears in *An eighteenth-century musical tour in central
Europe and the Netherlands* (ed. P. A. Scholes)
(*Dr. Burney's musical tours in Europe*, 2 vols., London,
1959), Vol. II, 201–203.

3   such terms as 'mathematical sciences' are 'routinely
used'   P. Gouk, *Music, science and natural magic in
seventeenth-century England*, Yale University Press,
New Haven and London (1999), 4.

5   the [Royal] Society's most overt interest in musical
subjects   P. Gouk, *Music, science and natural magic*, 62.

5   Newton in the mid-1660s 'learned all that had been
developed by modern mathematicians'   On Newton,
see P. Gouk, *Music, science and natural magic*, Chapter 7.

5   The work of Mersenne... representing 'a significant
milestone'   See P. Gouk, *Music, science and natural
magic*, Chapter 5, especially 170–8.

6   *Speculative is that kinde of musicke*   Thomas Morley,
Plaine and easie introduction to music, 'Annotations
necessary for the understanding of the Booke...'.

6   Music... belongs, as a science, to an interesting part of
natural philosophy   William Crotch, *Substance of
several courses of lectures on music, read in the University
of Oxford, and in the metropolis*, London, 1831
(reprinted, introduced by B. Rainbow, Clarabricken,
Co. Kilkenny, 1986), 1–2.

7   The science of music will not constitute the subject of
the present work   W. Crotch, *Substance of several
courses of lectures on music*, 3–13.

7   many were keen amateur musicians   A. Richards, *The
free fantasia*, p. 141.

7   the friendly atmosphere and liberal exchange of ideas
H.-G. Ottenberg, *Carl Philipp Emanuel Bach* (translated
by P. J. Whitmore), Oxford (1987), 152–3.

8   Proposal for establishing a Laboratory of Acoustics
University of Oxford Commission: Evidence
(Parliamentary Papers, 1881), Supplementary
Evidence, 374.

9   Among the prime examples... must be counted the
works of J. S. Bach   On Bach, an accessible source of
articles on such aspects as mirror canons and fugues,
number symbolism, and others is the *Oxford composer
companions   J. S. Bach* (ed. M. Boyd), Oxford, 1999.

9   At a distance of over 200 years, Hindemith's... *Ludus
tonalis*   For Hindemith, the *New Grove modern masters:
Bartók, Stravinsky, Hindemith*, London, 1984 (source of
the quoted extracts included here) gives particularly
clear and assimilable information.

9   Christiaan Huygens... expressed a wish that composers 'would not seek what is the most artificial'
Quoted in H. F. Cohen, *Quantifying music: the science
of music at the first stage of the Scientific Revolution,
1580–1650*, Dordrecht (1984), 225.

9   Their shared concern is essentially... the power of
music   See Cohen, *Quantifying music*, preface,
pp. xi–xii; and, on the relationship between
'cosmology, music and poetry', and between science
and music, J. Hollander, *The untuning of the sky: ideas
of music in English poetry 1500–1700*, Princeton, 1961;
republished, 1993.

Chapter 1: Tuning and temperament: closing the spiral

The idea of the mathematical group structure lying behind scale construction is explored in Chapter 23, 'Groups and music', of F. J. Budden's *The fascination of groups*, Cambridge University Press, 1972.

Some of the ideas in this chapter are extended in C. Scriba's article (in Danish) 'Matematisk og musik' in *Normat* **38**(1) (1990), 3–17. The ideas of irrationality versus rationality are explored in R. Osserman's article 'Rational and irrational: music and mathematics' in *Essays in humanistic mathematics* (ed. Alvin M. White), published by the Mathematical Association of America in 1993. A standard text, albeit somewhat dry, is *Music, physics and engineering* by Harry F. Olson, published by Dover Publications.

A pleasantly informal approach to some of the ideas appears in Chapter 11, 'Numbers, point and counterpoint', of *3.1416 and all that* by Philip J. Davis and William G. Chinn, Birkhäuser, 1985. Newton's ideas are comprehensively examined by Penelope Gouk in Chapter 5, 'The harmonic roots of Newtonian science', of *Let Newton be!*, edited by J. Fauvel *et al.*, Oxford University Press, 1988. Readers of French will enjoy two articles in *Kleisleriana* (Cahier du groupe mathématique et musique), No. 1, IREM de Basse-Normandie, 1985: 'Histoire du temperament' by B. Hacquier and 'Gammes naturelles' by Y. Hellegouarch.

Donald Parker explores the number-theoretic side in more detail, though with a pleasantly light touch, in 'Number harmony', *Mathematical intelligencer* **8** (4), 1986.

Chapter 2: Musical cosmology: Kepler and his readers

30 Recent research has shown   B. R. Goldstein, 'The Arabic version of Ptolemy's planetary hypotheses', *Transactions of the American philosophical society* **57** (1967), 3–16.

30 *Secret of the Universe*   Johannes Kepler, *Mysterium cosmographicum*, Tübingen, 1596; see also *Mysterium cosmographicum. The secret of the universe*, second edition, Tübingen, 1621: reprinted with translation by A. M. Duncan and introduction and commentary by A. M. Duncan and E. J. Aiton, New York, 1981.

30 absurd and monstrous   *Mysterium cosmographicum*, Chapter 16.

31 Lengths corresponding to the standard consonances Johannes Kepler, *Mysterium cosmographicum*, Chapter XII. A more detailed account of Kepler's theory is given in J. V. Field, *Kepler's geometrical cosmology*, London and Chicago, 1988.

31 Kepler decided that it was based on a corrupt text See Andrew Barker, *Greek musical writings*, 2 vols. (I: *The musician and his art*; II: *Harmonic and acoustic theory*), Cambridge, 1984, 1989; the second volume contains a translation of Ptolemy's *Harmonica*.

32 However, Kepler has departed   Details of Kepler's astrological theories are given in J. V. Field, 'A Lutheran astrologer: Johannes Kepler', *Archive for history of exact sciences*, **31**(3) (1984), 189–272.

33 his lack of success is displayed in the form of two tables   The first two tables in Chapter IV of Book V, J. Kepler, *Harmonices mundi libri V*, Linz, 1619. See also *Johannes Kepler. Five books of the Harmony of the World*, translation, introduction and notes by E. J. Aiton, A. M. Duncan and J. V. Field, *Transactions of the American philosophical society,* **209**, Philadelphia, 1997.

35 A particularly spectacular set   See J. V. Field, *Kepler's geometrical cosmology*, London and Chicago (1988).

36 When Kepler objected that no astronomer   See J. V. Field, 'Kepler's rejection of numerology', in *Occult and scientific mentalities in the Renaissance* (ed. B. W. Vickers), Cambridge (1984), 273–96.

38 Figure 5: Mersenne frontispiece   Marin Mersenne, *Harmonie universelle*, Paris, 1636; 'facsimile' reprint (reduced size), 3 vols., Édition du Centre National de la Recherche Scientifique, Paris (1963).

39 Actually, it goes back to the sixteenth   See W. B. Ashworth, 'The persistent beast: recurring images in early zoological illustration', in *The natural sciences and the arts* (ed. Allan Ellenius), *Acta universitatis Upsaliensis, Figura Nova* **22**, Almquist & Wiksell, Uppsala (1985), 46–66.

43 Kepler's work is important   See J. V. Field, 'Kepler's cosmological theories: their agreement with observation', *Quarterly journal of the Royal astronomical society* **23** (1982), 556–68.

Discography   Music by some of the composers whose names are to be found in accounts of music at the Court of Rudolph II in Prague, in particular pieces by Camillo Zanotti (c.1545–91), who is known to have worked for one of Kepler's patrons, is to be found on Capella Rudolphina, Duodena cantitans, and Michael Consort, conductor Petr Daněk, *Musica temporis* Rudolphi II, Supraphon, (1994), CD 11 2176–2231.

Chapter 3: The science of musical sound

Most of the material discussed in this chapter is discussed in greater detail in the author's *Exploring music*, published by the Institute of Physics in 1992.

Chapter 4: Faggot's fretful fiasco

61 Today's Western music See Ottó Károlyi, *Introducing Music*, Penguin (1965), for a musician's introduction to the principles behind Western musical scales.

62 *Frankie and Johnny* The origins of this song are not known, but its structure suggests that it is probably from the Mississippi valley in the pre-blues period of the 1890s; a version, which includes the chord sequence, can be found in Alan Lomax (ed.), *The Penguin book of American folk songs*, Penguin Books (1964), 121.

63 in order to create a harmonious scale See C. A. Taylor, *The physics of musical sounds*, Edinburgh University Press (1965), for a mathematical introduction to the principles behind Western musical scales; see also the Notes on Chapter 1.

65 The transcendence of π For proofs that e and π are transcendental, see Ian Stewart, *Galois theory*, Chapman and Hall, London (1989).

66 Eutocius, a commentator from the 6th century AD See Ivor Thomas, *Selections illustrating the history of Greek mathematics* (2 vols.), Heinemann, London (1939); this book contains information on many special geometrical problems, including angle trisection by neusis construction, conic sections, the quadratrix, and other methods.

66 David Fowler argues that See David Fowler, *The mathematics of Plato's Academy: a new reconstruction*, Clarendon Press, Oxford, 1987; this book contains a wealth of material on the relationship between continued fractions and early Greek mathematics.

67 Duplicating the cube amounts to solving For the impossibility of duplicating the cube, see Ian Stewart, *Galois theory*, Chapman and Hall, London (1989).

68 In 1581 Vincenzo Galilei See Vincenzo Galilei, *Dialogo della musica antica e moderna*, Florence (1581), 49; see also J. M. Barbour, *Tuning and temperament*, Michigan State College Press (1951; 2nd ed., 1953).

68 In 1636 Marin Mersenne See Chapter 2 and Marin Mersenne, *Harmonie universelle*, Paris (1636), 68.

68 In 1743 Daniel Strähle Daniel P. Strähle, 'Nytt påfund, til at finna temperaturen i stämningen för thonerne på claveret ock dylika instrumenter', *Proceedings of the Swedish academy* **IV** (1743), 281–6.

68 The geometer and economist Jacob Faggot Jacob Faggot, 'Trigonometrisk uträkning, på den nya temperaturen för theonernes stämming å claveret', *Proceedings of the Swedish academy* **IV** (1743), 286–91.

71 It was not until 1957 that J. M. Barbour J. M. Barbour, 'A geometrical approximation to the roots of numbers', *American mathematical monthly* **64** (1957) 1–9.

73 Isaac Schoenberg did the same in 1982 Isaac J. Schoenberg, 'On the location of the frets on a guitar', *American mathematical monthly* **83** (1976) 550–2, and *Mathematical time exposures*, Mathematical Association of America (1982).

73 the most natural thing to do This approach was first published in Ian Stewart, 'Les mathématiques de la gamme musicale', *Pour la science* **151** (May 1990), 108–14, and reprinted in English in Ian Stewart, *Another fine math you've got me into...*, W. H. Freeman, New York (1992).

74 a beautiful theory of the so-called *Pell equation* For further details, see L. J. Mordell, *Diophantine equations*, Academic Press, New York (1969).

75 Strähle's function is then obtained In fact, as David Fowler has pointed out, while $\frac{12}{17}$ is not a convergent of the continued fraction for $\sqrt{2}$, it is a so-called *intermediate convergent*.

Chapter 5: Helmholtz: combinational tones and consonance

77 Helmholtz's book: *Die Lehre von den Tonempfindungen als physiologische Grundlage für die Theorie der Musik* (1st ed. 1863, 4th ed. 1877); translated as *On the sensations of tone* by A. J. Ellis (1875, 2nd ed. 1885), reprinted by Dover, New York (1954); all page-references here are to the Dover edition.

77 Bosanquet's enharmonic harmonium This instrument, constructed in 1876, is in the Science Museum, London; for details of its operation and use, see Ellis's translation of Helmholtz's book, pp. 427–30, 479–81.

77 For a summary of Helmholtz's life (1821–94) and work, with bibliographies, see the introduction to the Dover edition and the entry by R. S. Turner in the *Dictionary of scientific biography*.

77 On A. J. Ellis (1814–90), see the *Dictionary of national biography*, Vol. 22. His further passionate interests in etymology, phonetics and pronunciation shine through this translation; see, for example, his long note on p. 24 on the appropriate renderings of the German *Ton* and *Klang*, from which the following is but one sentence: '*Timbre*, properly a kettledrum, then a helmet, then the coat of arms surmounted with a helmet, then the official stamp bearing that coat of arms (now used in France for a postage label), and then the mark which declared a thing to be what it pretended to be, Burns's 'guinea's stamp,' is a foreign

word, often odiously mispronounced, and not worth preserving.' He was a friend of J. A. H. Murray, the founding editor of the *Oxford English dictionary*; there are many details about him in Murray's affectionate biography by his granddaughter K. M. E. Murray, *Caught in the web of words*, Yale University Press, 1997.

78   These tones are heard   Helmholtz, pp. 152–3.

80   principle of conservation of energy   See, for example, T. S. Kuhn, 'Energy conservation as an example of simultaneous discovery', in M. Claggett (ed), *Critical problems in the history of science*, University of Wisconsin Press (1959), 321–56.

80   The mechanical problem   Helmholtz, p. 134.

81   *Ohm's law of perception*   Helmholtz, pp. 33, 56.

81   If, then, we assume   Helmholtz, p. 413.

81   History abounds with unwarranted rejection *and* Foucault presented the results   'A late-twentieth century resolution of a mid-nineteenth century dilemma generated by the eighteenth-century experiments of Ernst Chladni on the dynamics of rods', *Archive for the history of the exact sciences* **43** (1991), 251–73, on p. 255.

82   Hermann von Helmholtz   This photograph appears as the frontispiece of J. G. McKendrick's Hermann Ludwig Ferdinand von Helmholtz, Fisher Unwin, London (1899).

82   the Pythagorean association of consonance   For the Greek texts, with commentaries, see A. Barker, *Greek musical writings*, Vol. 2, Cambridge University Press (1989); the quoted texts below come from pp. 55–6, 191–3 & 160. For a scholarly assessment of Pythagoreanism, see W. Burkert, *Lore and science in ancient Pythagoreanism*, Harvard University Press (1972).

84   The problem of explaining consonance was a live issue   See H. F. Cohen, *Quantifying music*, Reidel (1984); the quotations from Kepler and Galileo are on p. 11.

85   consonance is a continuous . . . sensation of tone   Helmholtz, p. 226.

85   in a celebrated prediction of Helmholtz   Helmholtz, p. 211.

86   Helmholtz took the simplest such kind . . .   Helmholtz, p. 417.

86   knowing that diagrams . . .   Helmholtz, pp. 192–3.

87   I do not hesitate   Helmholtz, p. 227.

87   for example, the elaborate connection . . .   Helmholtz, pp. 422–30, 470–83.

CHAPTER 6: THE GEOMETRY OF MUSIC

A good introduction to symmetry in general (particularly in nature and the visual arts) is H. Weyl, *Symmetry*, Princeton University Press (1952). After this, one can browse through a textbook of geometry. Three recommended books are H. S. M. Coxeter, *Introduction to geometry*, John Wiley & Sons, New York (1969); P. M. Neumann, G. A. Stoy and E. C. Thompson, *Groups and geometry*, Oxford University Press (1994); and D. A. Brannan, M. F. Esplen and J. J. Gray, *Geometry*, Cambridge University Press (1999).

CHAPTER 7: RINGING THE CHANGES: BELLS AND MATHEMATICS

Dorothy L. Sayers' detective story *The nine tailors*, Gollancz (1934), is the most exciting introduction to change ringing.

Extensive information about the history and lore of change ringing is available in *Change ringing: the history of an English art* (general ed. J. Sanderson), The central council of change bell ringers, Vol. 1 (1987), Vol. 2 (1992) and Vol. 3 (1994); John Camp, *In praise of bells*, Robert Hale, London (1988); Ron Johnson, *Bellringing*, Viking (1986); and Wilfrid G. Wilson, *Change ringing*, Faber & Faber, London (1965).

Simple mathematical articles about change ringing include Arthur White and Robin Wilson, 'The hunting group', *Mathematical gazette* **79** (1995), 5–16, and B. D. Price, 'Mathematical groups in campanology', *Mathematical gazette* **53** (1969), 129–33. More advanced mathematical papers may be found in the papers listed in the Notes for this chapter.

Further information about the mathematics of change-ringing can be found in the following papers:

Deryn Griffiths, 'Twin bob compositions of Stedman triples', *Bulletin of the institute of combinatorics and its applications* **16** (1966), 65–76;

R. A. Rankin, 'A campanological problem in group theory', *Mathematical proceedings of the Cambridge philosophical society* **44** (1948), 17–25;

W. H. A. Thompson, *A note on Grandsire Triples*, London, 1886 (reprinted in W. Snowdon, *Grandsire*, London, 1905: revision of J. Snowden, *Grandsire*, 1888);

Arthur T. White, 'Ringing the changes', *Mathematical proceedings of the Cambridge philosophical society* **94** (1983), 203–15;

Arthur T. White, 'Ringing the cosets', *American mathematical monthly* **94** (1987), 721–46;

Arthur T. White, 'Ringing the cosets II', *Mathematical proceedings of the Cambridge philosophical society* **105** (1989), 53–65;

Arthur T. White, 'Fabian Stedman: the first group theorist', *American mathematical monthly* **103** (1996), 771–8.

129 Since 1991, C. Curtis-Smith's *Concerto for left hand and orchestra* has received performances in Detroit, New York and Tokyo.

CHAPTER 8: COMPOSING WITH NUMBERS: SETS, ROWS AND MAGIC SQUARES

Excellent introductions to issues in twentieth-century music, including some technical discussion of works, can be found in Paul Griffiths, *Modern music and after: directions since 1945*, Clarendon Press, Oxford (1995), and Arnold Whittall, *Musical composition in the twentieth century*, Oxford University Press, Oxford (1999).

A concise but comprehensive introduction to the twelve-note compositional techniques of Schoenberg, Berg and Webern is given in George Perle, *Serial composition and atonality*, 5th edn., University of California Press, Berkeley (1981).

One of the most highly developed accounts of the creative possibilities offered by mathematics for music is to be found in Iannis Xenakis, *Formalized music. Thought and mathematics in composition*, Pendragon Press, Stuyvesant, New York (revised in 1992).

Controversial and thought-provoking accounts of the structure of the music of Bartók and Debussy in terms of golden section and Fibonacci numbers can be found in Ernö Lendvai, *Bela Bartók: an analysis of his music*, Kahn & Averill, London (1971), and Roy Howat, *Debussy in proportion. A musical analysis*, Cambridge University Press, Cambridge (1983).

132 In music there is no form *and* The introduction of my method   Arnold Schoenberg, 'Composition with twelve tones', *Style and idea* (ed. L. Stein, tr. L. Black), Faber & Faber, London (revised 1984), 244 and 223–4.

133 For the rest   Anton Webern, *The path to the new music* (ed. W. Reich, tr. L. Black), Universal Edition, London (1975), 53.

134 For a detailed discussion of the *Lyric Suite*, see G. Perle, 'The secret programme of the Lyric Suite', *Musical times* **118** (Aug–Oct 1977), 629–32, 709–13, 809–13.

135 It has also, my Hanna   G. Perle, 'The secret programme of the *Lyric Suite*', *Musical times* **118** (Aug–Oct 1977), 709.

135 its 'suitability for study': Arnold Whittall, *Music since the First World War*, Dent, London (1977), 174.

137 An exhaustive analysis of the serial organisation of *Structure Ia* is to be found in György Ligeti, 'Pierre Boulez: decisions and automatism in *Structure Ia*', *Die Reihe* 4 (English edn. 1960), 36–62.

139 composition and organisation cannot be confused   Pierre Boulez, *Stocktakings from an apprenticeship* (tr. Stephen Walsh), Clarendon Press, Oxford (1991), 16.

139 in matters of rhythmic style   Paul Griffiths, *Peter Maxwell Davies*, Robson, London (1982), 25.

139 'projected' through the magic square   Quoted in Paul Griffiths, *Peter Maxwell Davies*, Robson, London (1982), 74; Griffiths gives a detailed account of *Ave maris stella* on pages 72–9 of this book.

139 by which a base metal may be transformed *and* governs the whole structure   Quoted in Paul Griffiths, *Peter Maxwell Davies*, Robson, London (1982), 163–5.

140 sequences of pitches and rhythmic lengths *and* are a gift to composers   Paul Griffiths, *Peter Maxwell Davies*, Robson, London (1982), 164, 173.

145 I discovered on coming into contact *and* With Le Corbusier I discovered architecture   Xenakis (1977), quoted in Nouritza Matossian, *Xenakis*, Kahn and Averill, London (1986), 53, 55.

145 possible to produce ruled surfaces   Iannis Xenakis, *Formalized music. Thought and mathematics in composition*, Pendragon Press, Stuyvesant, New York (revised 1992), 10; he goes on to demonstrate 'the causal chain of ideas which led me to formulate the architecture of the Philips Pavilion from the score of *Metastasis*'.

145 sound events made out of a large number   Xenakis (1972), quoted in Nouritza Matossian, *Xenakis*, Kahn and Averill, London (1986), 58.

146 Xenakis's symbolic music   Paul Griffiths, 'Xenakis: logic and disorder', *Musical times* **116** (April 1975), 330.

146 he gives us something only an artist can give   Nouritza Matossian, *Xenakis*, Kahn and Averill, London (1986), 243–4.

146 the effort to make 'art' while 'geometrizing'   Iannis Xenakis, *Formalized music. Thought and mathematics in composition*, Pendragon Press, Stuyvesant, New York (revised 1992), *ix*.

CHAPTER 9: MICROTONES AND PROJECTIVE PLANES

A historical account of various equally tempered systems appears in J. M. Barbour's *Tuning and temperament: a historical survey*, published by Michigan State College Press (1953).

An introduction to atonal music and the mathematical tools it uses can be found in A. Forte's *The structure of atonal music*, Yale University Press (1973), and J. Rahn's *Basic atonal theory*, Longman (1980).

The relationship between transposition and inversion, among numerous other foundational matters, is formalized in D. Lewin's *Generalized musical intervals and transformations*, Yale University Press (1987).

The compositional employment of pitch classes receives a thorough explication in R. D. Morris's *Composition with pitch-classes*, Yale University Press (1987).

The history of major–minor dualism is traced in D. Harrison's *Harmonic function in chromatic music*, University of Chicago Press (1994).

Finally, a fuller discussion of block designs, projective planes and difference sets can be found in I. Anderson's *A first course in combinatorial mathematics*, 2nd edn., Oxford (1989),

or J. H. van Lint and R. M. Wilson's *A course in combinatorics*, 2nd ed, Cambridge University Press (2001).

Chapter 10: Composing with fractals

163 A glance through the illustrations   See Heinz-Otto Peitgen and Dietmar Saupe (eds.), *The science of fractal images*, Springer, New York (1988), and Heinz-Otto Peitgen and Peter H. Richter, *The beauty of fractals*, Springer, New York (1986). Chaotic dynamical systems are described in Chapter 3, and the Mandelbrot set in Chapter 4, of Peitgen and Saupe's book.

163 A. K. Dewdney's mathematics column appeared in *Scientific American*, September 1986.

Overview

William Crotch - Courtesy of Norfolk Museums Services

Chapter 1

Britten – *Serenade* Reproduced by permission of Boosey & Hawkes Music Publishers Ltd.

Chapter 3

The opening image was supplied by Professor Taylor's family. The figures in this chapter appear as Figures 1.3, 1.5, 1.13, 1.16, 1.21, 2.2, 2.32, 3.12, 4.16, and 4.22 in Charles Taylor's *Exploring music*, Institute of Physics Publishing (1992).

Chapter 4

Figures 1–5, 10, and those in Boxes A and B appear as Figures 104–6, 110–12, 113a and 114 in Ian Stewart's *Another fine math you've got me into . . .*, W. H. Freeman, New York (1992). The opening image and Figures 6 and 7 appear in *Proceedings of the Swedish Academy* IV (1743), pp. 281, 289 and Table VI Figure 4. Figures 8 and 9 are from Ian Stewart.

Chapter 5

The picture of Bosanquet's enharmonic harmonium appears courtesy of The Science Museum / Science & Society Picture Library (picture number SOUC000006).

Chapter 6

Wilfrid Hodges is very grateful to Nicolas Hodges for wise advice and some good examples. The musical illustrations in the chapter were typeset using Finale 2000a.
Johann Jakob Froberger – *Suite XII in C major*
The music collection of the Oesterreichische Nationalbibliothek
Tippett – *A child of our time, No 26*
© 1944 Schott & Co. Ltd., London. Reproduced by permission.
Ives – *Duty*
© Merion Music Inc., King of Prussia, PA USA. Copyright renewed. International copyright secured. Reproduced by permission of Alfred A. Kalmus Ltd.
Kugel – *Translation-Rotation*

Chapter 7

Dermot Roaf and Arthur White are very grateful to Dr John Pusey, deputy Captain of the Ringers at St. Giles' Church, Oxford, for his advice and help with describing the musical aspects of change-ringing and for general constructive criticism of drafts of this chapter.

Chapter 8

Boulez – *Structures Ia*
©1955 by Universal Edition (London) Ltd., London. Reproduced by permission.
Peter Maxwell Davies – *Ave Maris Stella*
© Copyright 1976 by Boosey & Hawkes Music Publishers Ltd.
Reproduced by permission of Boosey & Hawkes Music Publishers Ltd.
Xenakis – *Metastasis*
© Copyright 1967 by Boosey & Hawkes Music Publishers Ltd.
Reproduced by permission of Boosey & Hawkes Music Publishers Ltd.

Chapter 9

Canto LXXXI (10-line excerpt) by Ezra Pound is taken from *The Cantos of Ezra Pound*, © 1948 by Ezra Pound; used by permission of New Directions Publishing Corporation.

Chapter 10

Robert Sherlaw Johnson–Courtesy of Decca Record Company Ltd. The editors are very grateful to Dr Robert Lockhart for reproducing the fractal images.